FROM POVERTY TO POVERTY

A Scotsman Encounters Canada

*Best Wishes from :-
Ian Moore-Morrans*

by Ian Moore-Morrans
edited by Gayle Moore-Morrans

DEDICATION

To Gayle, my beloved wife, with thanks for editing this book with such great dedication and for becoming more Scottish in the process!

To my daughters, Audrey German and Shirley Lee, who "lived" part of this story along with their mother, Mary, and me. My love and thanks to both of you.

Copyright © 2012 by Ian Moore-Morrans
First Edition – March 2012

Edited by Gayle Moore-Morrans

ISBN
978-1-77097-247-6 (Hardcover)
978-1-77097-248-3 (Paperback)
978-1-77097-249-0 (eBook)

All rights reserved.

No part of this publication may be reproduced in any form, or by any means, electronic or mechanical, including photocopying, recording, or any information browsing, storage, or retrieval system, without permission in writing from the publisher.

Published by:

FriesenPress
Suite 300 – 852 Fort Street
Victoria, BC, Canada V8W 1H8

www.friesenpress.com

Distributed to the trade by The Ingram Book Company

Library and Archives Canada Cataloguing in Publication

Moore-Morrans, Ian, 1932-
  From poverty to poverty : a Scotsman encounters
Canada / Ian Moore-Morrans ; editor, Gayle Irene
Moore-Morrans.

Includes bibliographical references.
Also issued in electronic format.
ISBN 978-1-77097-248-3 (pbk.).--ISBN 978-1-77097-247-6 (bound)

1. Moore-Morrans, Ian, 1932-. 2. Scottish Canadians--Biography.
3. Scots--Canada--Biography. 4. Immigrants--Canada--Biography.
5. Machinists--Canada--Biography. I. Title.

FC106.S3Z7 2012   971.004'91630092   C2011-908249-7

TABLE OF CONTENTS

Dedication iii

Preface vii

Chapter One: The Cold And Hungry Early Years 1

Chapter Two: During And Just After The Big War 29

Chapter Three: The Move To The "High Class Hovel" 61

Chapter Four: Off To The Royal Air Force 93

Chapter Five: The Stinking Egypt Stint 111

Chapter Six: Back To "Dear Old Blighty" 173

Chapter Seven: The Canadian Connection 225

Chapter Eight: Big Decision Coming Up:

    Canada Or Australia? 235

Chapter Nine: The Big Step

    (And It's Bigger Than You Think, Boy!) 243

Chapter Ten: Atlantic Ocean Crossing Ends In Collision 249

Chapter Eleven: Oh My, What Have I Done? 273

Chapter Twelve: We Get "Established" At Last 289

Chapter Thirteen: We Move To An Apartment

    And Then To A Rental House 307

Chapter Fourteen: At Last We've Made It! 327

Acknowledgements 344

About The Author And Editor 345

PREFACE

My principle reason for writing my autobiography is that I have met so many people on the Canadian side of the Atlantic Ocean whose backgrounds are Scottish, English, Welsh, Irish or whatever, who have no idea *who* their grandparents or great-grandparents were, what they did or how they lived. Thus I decided that my descendants, friends and even strangers should get to know me, if they so desire.

Several times I've found myself checking out through a grocery counter and spoken a few words to the clerk. Upon hearing my Scottish "burr" (folks over here think that Scots speak with a "brogue"—no, that's Irish!) she would invariably ask me if I were Scottish and then tell me that her grandfather (or grandmother) was Scottish. When I asked her where he or she was, she would then tell me the relative was dead. When I inquired where in Scotland they came from, she didn't know. She didn't know anything about him or her—and that happened more than once. On arriving home one day from a little bit of grocery shopping, I told my wife, "I'm going to write my life story for my descendants to read—they should know who and what their grandfather did while he was alive." So here is the first volume—all done and delivered!

This book may not always be chronologically correct. As I searched my memory, I wasn't always sure what things happened in the same time frame or exactly what age I was when something happened. However, nothing has been embellished or intentionally made brief. What I have written is exactly as I remember it, albeit sometimes a trifle fuzzy. I have occasionally changed a person's name to maintain their privacy. The only incorrect information might be a street name spelling or a slight error in a date; but be assured that all that is included really did happen. It is *my life*, written with the express intention of filling in information that will not be accessible after I am no longer living. I've tried to remember the good times, the bad times, the funny times and the sad times, from 1935 until 1970. The second volume ("Came to Canada, Eh?", as yet unpublished) will cover the years 1970-2004, maybe even later than that if I continue living and writing. My present wife and editor is urging me on to Book Three!

Most of this story was written when my name was still "Ian Morrans."(The name on my birth certificate is "John Morrans" but I was always called "Ian," the Scottish Gaelic equivalent of "John." Sometime after I immigrated to Canada, I legally changed my first name to "Ian.") My name change to "Moore-Morrans" happened in 2003 when I remarried after I lost Mary, my first wife (who was living when I wrote most of the first two volumes of this autobiography). The maiden name of Gayle, my new wife, is "Moore" and she wanted me to add it to my family name, which I gladly did. This was a bit of a coincidence as my stepfather's name was similar—"Moorhead." I suspect all three names (Morrans, Moore and Moorhead) have an ancient common root in the Celtic languages.

~ * ~

Around 2000, I was contacted by Eric Morrans, a distant cousin in Scotland, who was in the process of compiling a family history. According to his research, the name "Morrans" was a recent spelling. Apparently, each generation had a different spelling of the family name. Thus, in my family, "Morrins" in the early 19th Century became "Morrens" or "Mornes" which became "Morrns" (my grandfather) and somehow my grandmother and mother used the spelling as we knew it – "Morrans."

Perhaps a lot of them were illiterate or the record keepers were careless spellers!

~ * ~

The reader will notice that I've used the Scottish vernacular when Scots are speaking amongst themselves and normal English when they are speaking with non-Scots. That reflects my own speaking pattern. When among Scots, my speech becomes increasingly "Scottish-sounding," For example, "Ah" (I); "tae" (to); "ye" (you); "no" (not); "canna" (cannot); "oor" (our); "widna" (would not); "aboot" (about), "aye" (yes), etc. I've also used British words for the period before I emigrated from Scotland (such as "lorry," "chap," "bloke," and "cheerio"); and changed them to North American words sometime after I immigrated to Canada (such as "truck," and "guy").

My editor and I had a disagreement about allowing "Scottishisms" (as she calls them) into my narrative. I insisted on leaving them in, however, for that's how we (Scots) speak. Thus you'll find the occasional sentence such as "So, there's me, the great boatbuilder." instead of her "cleaned up" version, "So, there I was, the great boatbuilder."

* ~ * ~ *

Lochend part of Campbeltown, taken from the pier

CHAPTER ONE

The Cold and Hungry Early Years

Thinking back on my early childhood, the most miserable over-all time was when it was evening, dark outside, middle of winter, clothes damp and cold from the rain, no oil for the lamp, no candles either, which meant no light of any kind in the dump we lived in, no fire to warm me a wee bit and no food. As a wee boy of six years of age, I was just sitting all alone in a dingy one-room flat. My hair was wet and water was still running down my face from the rain. I was shivering and my teeth were chattering, as I sat hoping that maybe someone would come and light the oil lamp. Maybe that someone would have a few lumps of coal and there would finally be a nice, warm fire started. If I got too hungry I could always fill my belly with cold water; then I didn't feel so bad any more—well, for maybe an hour.

~ * ~

Campbeltown, on the Kintyre peninsula, Argyllshire, west coast of Scotland, was a picturesque little town of about four thousand people. That was all it had going for it, mind you—not much else. This was the time of the Great Depression and few people had jobs. There was very little work available except in the fishing industry.

In this setting lived my grandmother, my mother, my brother and me. There never was a dad. (There was a stepfather later but that's a different bit of the story.) My mother's surname was Morrans but Granny's name was Margaret Killin. I know that she was a "Mrs.," but I was never told of a grandfather. (Maybe he was another one of my family who was drowned at sea, like "Uncle Charlie" who was mentioned from time to time, or maybe he was the type who went out one day long ago for a packet of cigarettes and kept on going!) I assume Granny was married twice—first to a man named "Morrans" and then to a man named "Killin," (a worthy, old Scottish name). It was impossible to talk to Mother or Granny about any of this—ever! Any inquiries later on in life were met with, "Never mind asking questions." When I got older, the situation didn't bother me anymore, as I told myself that I had nothing to do with it; I was just the end result. To this day, I'm not sure whether my brother was a full or a half brother or whether my mother lived common law with someone who eventually took off. In the end, it didn't matter very much one way or another to me, and I doubt such a complex issue ever crossed my brother's mind! All this meant that there was no breadwinner in the family, nobody to go out and earn a living.

~ * ~

1935 was just about when life started for me, for I can't remember anything that happened in the three years of my life up to that point. That is most likely very fortunate, for we were, without doubt, the poorest family in that little town. If I'm mistaken then I certainly pity the other poor souls! No, I'm *certain* that we were the poorest, by far; for no one else in our town lived in such pathetic conditions as we did.

Yes, we were destitute! There never was the privilege of being able to say something like, "Oh dear me, we've run out of corn flakes!" Nobody thought of complaining—this was life—just exactly what we were used to, for we didn't know any better.

To put it bluntly, many times, quite often for more than two days at a time, we had nothing at all to eat in the house—not one thing—so I had to fend for myself. More than once I had to go to my friends' homes and ask their mothers, "Can ye gie me a piece an' jam, please?" (meaning, "May I have a slice of bread and jam?") I was never refused, although they often didn't have much more than we had; but I was too young to realize this at the time. I grew up constantly hungry. (Only after I joined the Royal Air Force at age 18 did I discover what it was like to have a full belly of half-decent food.)

The weather in Scotland is never what you could call "warm." When it was damp—which it usually was—especially in the evenings or when it was "smirring" (Scotch mist – a very fine rain) it seemed to go right through you. If there was a fire, you had to crowd close to it for ages before any warmth penetrated your body. It took money to buy coal and there was always an extreme shortage of money. Coal was needed not only to heat our place but also for cooking, which was done solely in a small fireplace which we called a "grate" (about nine inches square). If Mother had any money on hand, she had to make a big decision – buy some coal or some other necessity like oil for the lamp, or buy a little food. Usually it was food that won out. However, kerosene wasn't very expensive, so sometimes it was better to be hungry and to have a bit of light instead of sitting in the dark.

~ * ~

Where do you start when you are trying to tell the first part of your life? Earliest memories, obviously. I would most likely have been three or four years old at the time of the "First Great Memory" which involved my brother Archie who was exactly two years older than I. (Both of us were born on a May 2nd—he in 1930 and I in 1932). Everyone started school at five years of age in Scotland (no time-wasting, enjoy-yourself-thing like kindergarten for us!). It was a stern "you sit down there,

Lad, and learn" sort of thing. Straight into grade one—only we didn't have any grades. Archie had just started school. I remember him putting his new school books in a drawer and telling me, "Don't ye touch thae books." Then he went out. They were two picture books. Despite his warning, I know I took the books out of the drawer to have a good look at them. However, I cannot remember if I put them back again—nor can I remember the result of my disobedience. (If things went true to form, my brother probably found out and hit me.) The whole thing maybe lasted about ten seconds! Fancy telling a wee boy of three or four years of age not to look at something that would have appeared fascinating and unusual! That is *definitely* telling him to look. I wonder what would have happened if he hadn't said anything at all!

The next memory took place two full years later. I was being taken to my first day in Dalintober Elementary School. I can remember getting coloured paper to play with and this was the day that I discovered *MILK*. A tiny round glass bottle of white stuff, maybe about five inches high (12 cm.) and two inches wide (5 cm.), was given to each child at play time. We drank it straight from the bottle. The first part I drank was absolutely wonderful and it was a different colour. I learned later that it was the cream that had separated and settled on top of the milk. (We don't get milk like that any more; they say it isn't good for you!) The teacher later taught us to shake the bottle first to mix the milk and cream together and then to remove the little cardboard seal on the top. (I had pushed mine right into the bottle as I didn't know to pull on the little tab.) I was five years of age and had never tasted milk before. Was it ever good! I couldn't believe that all of it was for me. I know I didn't ever have any objection to heading for school each morning just to get some milk. No doubt I would have gone there Saturdays and Sundays too, if I could have!

Another milestone in my life was learning to tie my shoes; and I have a gypsy girl to thank for it. The gypsies in

Campbeltown were called "tinkers." We were told to have nothing to do with them. My memory tells me that others treated them as dirty and undesirable, something like lepers. I don't know why we were taught to look down on them, except perhaps that they were "different." Most people were every bit as poor as they were, and my family was definitely even poorer. I found them fascinating, for a tinker family had a pony and a cart so they could carry their possessions around with them no matter where they went. There was no way we could have afforded anything like that. We certainly didn't have *anything* worthwhile to put *in* a cart for starters, and it would have been hard for us to be able to afford a bit of horse *meat* if it had been available, let alone own a whole horse! The only thing we had that they didn't was a permanent roof over our heads, humble as it was and (maybe once a month) a sponge bath of sorts with water that barely had the chill taken from it—whether we needed it or not! But then again, I had no idea how the tinkers got their baths. Maybe they were cleaner than I was, for that wouldn't be difficult. And just think, even if we did happen to have a stick deodorant in those days (which we didn't), we wouldn't have been able to afford one anyway. We really must have stunk to high heaven!

Anyway, a kind gypsy-girl took pity on poor, ignorant wee Ian and taught me how to tie my shoe laces during a midmorning break my second year of school. The trouble, you see, is that I am left-handed. She sat facing me, so I was seeing it done backwards from where I was sitting. For the next 45 years or so I tied my laces opposite from everybody else.

I had been tying a granny knot and the bows of my laces laid at forty-five degrees to my shoes. One day I tied a reef knot. Then, hey; what do you know? As soon as I did so my laces lay *straight* across my shoes, just the same as everybody else's! (I probably shouldn't admit that I must have been about 50 years of age by that time!)

~ * ~

One extraordinary situation sticks out and makes me cringe to this day! I believe that Jewish boys are circumcised after they are seven days old—or thereabouts. Well, I was circumcised when I was *six years of age*! I can remember standing by our bed and Nurse Cameron "doing my willie" as I stood there! And, surprisingly enough, I can't even remember if it hurt in any way. I also have no idea if my mother "prepared me" in any way for the occasion or why this was done at such a late age. (More questions that will never be answered!)

~ * ~

It's time to describe the "home" I grew up in. I can't call our place either a "house" or a "flat." The dump we lived in was referred to as a "single end." I think the closest equivalent today would be "bachelor pad;" although I would use the comparison very loosely. It meant we had no kitchen, no living room, and no bedroom—not even a toilet; they were all wrapped into one—minus, fortunately, the toilet! I think that our rent was a half-crown (two shillings and sixpence, or about 36 cents) a week. This "home" consisted of one room about 10 feet square and was actually an attic; the outside wall rose about two feet and then sloped backward as it was then the roof. That was it! Out of what would have been classified as "undesirable living accommodations" in our town, our building was the only one that was eventually torn down. The others were renovated into habitable living quarters. Incidentally, there are *no* slums in Campbeltown today. Every place has been renovated and brought up to standard.

Everything we had was on the verge of being useless—otherwise whoever it belonged to originally would have kept it and we would have had absolutely nothing. Nothing we ever owned was paid for by us: first of all because whatever we had wasn't worth *anything* and secondly, if it *was* worth something then we wouldn't have had it—simple as that. Our total furniture consisted of one double bed, a six-drawer dresser, two wooden chairs and an ancient wooden kitchen table. That was it! Even today I can still picture it—"not enough room to swing a cat in" would have been a good description!

The four of us slept in the one double bed. My mother and grandmother slept at the top of the bed and my brother and I were at the bottom—with toes in the middle. If ever any bits of newspaper were "discovered" anywhere, they had to be taken immediately home—they were a "precious commodity." If the newspaper sheets were whole, they were put between our only two blankets to help to keep us warm until they eventually fell apart. (It is amazing just how much heat a few layers of newspaper add to the bed-clothes.) All of us slept on the mattress, and I mean *right on* the mattress, with nothing in between! I shudder now to think what state that mattress was in and where it had been in past years for it was also an outcast—something that nobody else had wanted! We didn't *ever* have bed sheets. (I didn't know that such things existed until I later went to the Royal Air Force and had sheets issued to me! When I first got them, I didn't know what to do with them!)

Our dresser had at one time included an attached mirror. That was long gone. What we used for a mirror was a fragment of what must have been the centre portion of a much larger mirror. It was about two square feet in area, (*not* two feet square, which is twice the size), was quite irregular in shape, sort of triangular, and sat on the dresser on two of the three broken points with the other part of the triangle leaning against the wall. That was okay, for there was no fear regarding the broken edges of the mirror scratching the finish on the dresser which was already completely chipped and scratched. As the handles were missing, we opened the drawers by pulling on old rusty screw-nails inserted where the handles should have been. One of the dresser's legs was broken off and was supported from falling over by half a brick.

The table was just plain unpainted wood, with a wee bit of varnish left on the legs. It had a shallow drawer in the middle where we kept some junk utensils. I can remember clearly that the "cutlery" included four sort-of bone-handled knives. (Actually, I think they were "butter knives."). They had only half a handle each as the other half had broken off, which is probably why *we* now had them. They were so dull you couldn't cut yourself even if you tried.

Our two old kitchen chairs both "rocked," which means that the legs were either of different lengths or the floor was uneven—probably both. When my mother and Granny were at home, my brother and I had to sit on the floor or on the edge of the bed. Sitting on the floor got us closer to the fire when we were lucky enough to have one. The floor was ugly, very dirty linoleum that hadn't seen water for many years, and there was absolutely no carpeting whatsoever. I don't think that anybody I knew had *any* carpeting of any kind in their home, not even a throw rug.

Two old brass candlesticks decorated the mantle shelf above the fireplace. There were two ancient pots—one big one and one smaller, an ugly cast iron fry pan, a decrepit kettle and a chipped and cracked teapot. Every one of them was crusted black from countless years of sitting on top of a coal fire.

Oh, I nearly forgot—we had a poker. The real necessities of life back then were air, water, food and a poker. Everyone had to have one, as everyone had a fireplace. The fireplaces were very inefficient, for most of the heat went up the "lum" (Scottish for "chimney").

What I've just listed was the total count of *all* our worldly possessions, except for maybe a few "odds and sods." Oh yes, sorry—we also had an old alarm clock which sat between the two candlesticks. The glass that should have been on the clock's face was missing, as was one of the two bells on top. There was an oil stain on half the face, showing that someone in the past had been a bit too ambitious with an oilcan.

Not one picture was to be found on our only true wall (remember, we lived in an attic), nor anything resembling something that could have been used to "beautify" the place—no vase to hold flowers (that would have been a joke!), no knickknacks which usually are used to make a place "interesting." No, the place was bare; reflecting the fact that whoever lived there had next to nothing.

The wall probably didn't need decoration anyway as it had a very fine dusting of soot ingrained into it because of the many "blowbacks" from the fireplace, filling the place with smoke when we were lucky enough to have a fire. There wasn't really enough distance from the fire to the chimney pot to create a decent up-draught. (That's "updraft" on the North American side of the big pond.)

As far as light goes, there was no electricity, or even a gaslight fixture that used the special gas "mantles." No; we used the type of oil lamp with a glass chimney seen in old time western movies when we could afford the kerosene (we called it "paraffin oil"). Failing that, we used candles, but only one at a time. Sometimes we couldn't even afford a candle, so were forced to sit in the dark. When we were lucky enough to have a fire, we could watch the "fluffin' low." ("Low" rhymes with "how." It described the action of the flames as they left the hot coals in the fireplace.) In our entertainment-deprived environment, this was nearly as good as watching modern television, maybe even better, depending on your imagination. It was amazing just how many pictures could be seen in the flames, if only for a second!

On those awful nights with no "fluffin' low" to distract our thoughts, we just sat, cold, damp and hungry, until it was time to go to bed. The winter nights were very long but this was something we were used to. When the morning came, this meant there was to be no breakfast before going to school, for there was no money in the house to buy anything. No money for coal to heat the place, no money for oil for the lamp or for candles to give us some light and no money for food.

All after-dark activities were (usually) done by oil lamp or by candlelight. (Good job I didn't *ever* have any homework from school—honest!) This level of comfort also applied to the higher class dumps that were downstairs, as there weren't any of the above services in the building known as #8 Lorne Street, Campbeltown, Argyllshire, Scotland, which, incidentally, is no longer in existence, having been torn down during the 1950s (in my opinion, about fifty or more years too late)!

There was no running water inside our home. We had to go out to the landing to get water from an old, black cast iron sink (called a "jaw-box" by my granny—I've no idea why). The cold-only faucet was shared with another two "*bigger* and *better* dumps" than ours. (I'm saying "bigger," as one home had one tiny bedroom and the other had two; however, they were still dumps and still slums!) Besides the sink, this landing also included the toilet, which was shared with the other families. I don't know how often the toilet was washed or cleaned; all I can say is that I never saw it done; although I know that each place was supposed to take a turn doing so. I probably shouldn't complain about the shared toilet on the landing, especially when shortly after I came to Canada, I found out about the outhouses of rural North America and the awful smell that arose from those outhouses, especially with the summer heat. Wow, I couldn't believe it! In Scotland, we had never ever dreamed that such awful things existed (although I realized that they were absolutely necessary). It would have made us appreciate what we had; for compared to them, our toilets could be classed as "near Utopian" facilities! Our toilets had *running* water and *flushed*! Also, it wasn't too cold for us to sit on the throne in the middle of winter and one didn't have to battle blizzards when the call of nature arose. (A cold winter day for us would have been around +3 Celsius or 38 Fahrenheit). I later learned of homes that had an *extra* toilet just outside the main building, at the back of the house. This was in England and they were flush toilets, not "dry" or chemical. Possibly similar ones existed in some places in Scotland, but I was unaware of it.

Bits of newspaper also served us as toilet paper if we didn't need it for the bed. We certainly couldn't afford to even *buy* a newspaper. Any "found" newspaper would be taken to the toilet with us, a bit torn off and the remainder taken back into the house – not left for the neighbours to use—they didn't ever leave any for us! Rolls of toilet paper weren't invented yet—at least I don't think they were! Any other paper that was found was to be kept for trying to light the fire when we had coal. I've just remembered how difficult it was to light the fire. People used to buy "kindling"—small diameter wood that was used to get the fire started. This, to us, was a sheer luxury! Archie and I would go to the woods sometimes to bring back some small, dry branches for kindling. Often we would forget to get some and then it was very hard to get the fire going. Sometimes we would "steal" the paper from between the blankets, roll them up tight and "wring" them, hoping that they would burn long enough to get the coals started.

~ * ~

Our house was entered from the street by way of a "close." (Pronounce "close" as in "near", not like "close"—"to shut") This was an entryway a little over six feet (two meters) in height, maybe three-and-a-half feet (1.1 m.) wide and about 15 or so feet (five m.) deep. Our close resembled a man-made cave—not only at night, for even the *daylight* hardly penetrated to the far end! Half way along this passage was a door on either side for the ground floor residents (only an old lady in each one). At the end of this passage, facing you, was a door that took you out to "the back" which was an open area where peoples' clothes lines were strung, across the way, from the wall of our building to the opposite wall, which was the outside wall of McConnachie's garage, one after another, for all the different families (26 in all, I think) who lived on our street.

All of the houses on our street were of a bit better quality than ours was. And if you happened to get mixed up and used the wrong line, you found your laundry lying on the ground later in the day and someone else's laundry hanging where yours used to be.

There were no laundry facilities whatsoever. I can remember seeing some lower class homes in different parts of Glasgow that had really ancient stone wash-houses in an area behind their slums (called "the back green", though they had no grass!). And I've a funny feeling that a few such buildings existed in my hometown. When we were up the "dirty back", (at the bottom of Burnside Street), we were able to climb onto flat roofs, most likely the wash-house roofs for the Burnside Street houses!

These wash-houses had large cast iron open pots (boilers) that were filled with water (using a pail) and a fire was lit underneath to heat the water. (You used your own coal.) The clothes were then swished around with a long stick, the water was brought to a boiling point, and the whites were done first. Then the clothes were fished out using the same long stick, the fire underneath was allowed to die a little earlier, and then the coloureds were washed, all in the same water, of course. I figured back then, that anyone who used these wash-houses must be the gentry and must be rich, and have nice homes. (I found out later they weren't and they didn't; but they didn't have it as rough as we did!)

Also in the same wash-house was a large, hand-driven wringer and a wooden tub (same as the "rub-a-dub-dub, three men in a tub" sort) which had to be filled with cold water for rinsing. Each tenant had a specific laundry morning or afternoon in the week; if it wasn't used when it should have been, then the turn was lost until the following week. In that case, everyone in that family went dirty, for folks back then didn't have the extra clothes that people have today. However, we didn't even have the "luxury" of such conditions; our laundry was done in an old galvanized tub in front of the fire, using lukewarm water. Thus, our "whites" *weren't*! I think that my mother had to borrow the tub, for there was nowhere it could have been stored in our small place without it being in the way and me remembering it. Although there's also the possibility it was kept under the bed!)

~ * ~

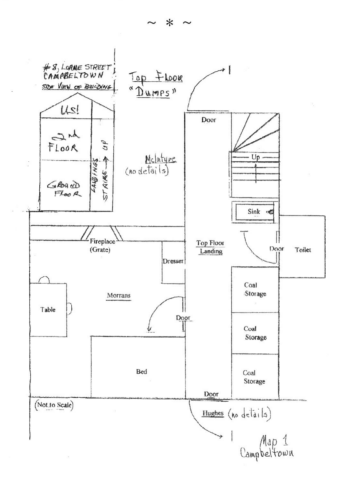

One of the "Top Floor Dumps," my first home

Back to the description of our "man-made cave." If you turned left at the end of the close, just before the door out to "the back," you then proceeded to climb the stairs. Up one flight was a small platform, which extended to the left, jutting out from the staircase. This was the first landing. Two better class homes were there, and they must have been a much better class (but still dumps), for these two homes covered the whole middle floor, making them both two bedroom flats. They didn't have inside toilets either, but they *did* have an inside sink with a *cold* water faucet.

Of course, the people in these "high class" dumps would never allow any of us from the low class dumps upstairs, into their homes, so I can't say what the interiors looked like! I can remember that one of the families was called Gilchrist. They were older, had no children at home and wouldn't talk to us; maybe they were *too* high class! I have no idea who the other people were, although I lived there for the first thirteen years of my life.

Continuing straight up the staircase (which had no means of lighting at night, not even an oil lamp), a "U" turn was made to get onto the top landing which ran along from the first door (McIntyre's), past the middle door (our family) to a third door (people called Hughes).

The two-bedroom unit at the first door was occupied by people named McIntyre from Rutherglen, a suburb of Glasgow, but only during the two months the children had off school in the summertime. I used to think that they were rich, as they had to pay the rent for the rest of the year when they weren't there. I realize now that the rent couldn't have cost very much as our building was, by far, the worst building in town. And I'm sure that no one else in town would have wanted to live there anyway! There never was a man in the family from Rutherglen, only a woman whom we *had* to call "Auntie Mary" (instead of Mrs. McIntyre) and two of her four children, Ian and Rita. The other two were old enough to work full time in Glasgow, so I guess that was why we very seldom saw them. Of course, the two who were "here" all summer were to be called our "cousins." (The relationship was more than I realized at the time as far as Rita was concerned. It wasn't clarified until I was seventeen.)

To get into our "palace" behind the middle door, you just put your thumb on the flat at the end of the "sneck bar" and pushed your thumb down. I think there are still similar "latches" on the market, used for barn doors, where the lever on the other side of the door is lifted out of a notch when the thumb is pushed down. There was neither a proper door handle nor any means of locking the door. Not that a lock was important, mind you. We didn't have anything worth stealing. *Not one thing.* (I think if a thief came into our place he would probably have felt sorry for us and would have left something!)

The area on the other side of the landing between the sink and the last (Hughes) door was divided into three sections, one for each "home", so that each could have a place to keep coal. Coal was expensive (for us), and when we could afford to buy any, it was only a half hundredweight at a time (56 pounds or 25 kg.). There was an unwritten law that nobody else touched your coal. It was important that heat was available in the damp, cold West Highland air, as the risk of tuberculosis was quite high at that time. In other words, it was all right to die of starvation but please don't die of TB! Central heating was unheard of in those days. Considering all the cold, damp, hungry nights I spent there, I often wonder why it is that I'm still alive! Seriously!

Our landing was at the rear of the building, up two flights, and ran at 90 degrees to the "close." Though I said earlier that there was no means of lighting the stairway, this wasn't entirely true. Both of the middle floor homes had a little shelf by the door where a small oil lamp sat just above head height, like an outside light. They were never lit unless the people there were having visitors or going out for an evening. (Hey, they wouldn't want to help the upstairs trash see their way in the dark!)

It was common practice to always have a box of matches hidden away somewhere dry so that at night we would be able to grope our way upstairs with the help of a little light. ("Safety" type matches hadn't been invented yet.) We couldn't carry the matches with us as they would catch fire in our pockets if we ran, as boys often did. This happened to me once and I was lucky that I only had about a half-dozen matches in the box. I ended up with no pocket, a slightly burnt thigh, but the outside cloth still looked alright, which was good. And as I had only one pair of pants (actually they were called "trousers", which we Scots pronounced "troosers" or "troos"), I had to make sure that they didn't catch on fire again, which is why I had to find a nice, dry place to hide the matches. That was only if I'd had the threehae'pence (three half pennies, or one-and-a-half pennies, about two cents) to buy them. A full box held 48 matches, so you can imagine half that amount would have made a beautiful blaze (miniature bomb), and probably would have sent me to hospital with a terrible burn.

Nowadays the solution would have been very simple; just fill the match box with cotton wool (absorbent cotton) or tissue so that there wouldn't be any looseness to allow them to rub against each other and ignite! Wee boys didn't think of things like that. Besides we didn't have any cotton wool anyway! Then, too, most times I had only one pair of troos. When they were getting washed (maybe once every six months!), I had to sit in our home on my bare bum while they dried in front of the fire, as we never *ever* (as boys) had *any* underpants to wear. They cost money. (Imagine a humorous picture – me pulling off flaming troosers in the middle of the street and wearing *no* underpants!) In those days boys didn't wear long trousers in Britain until school-leaving age. Up to age fourteen, we boys wore knee-length troosers, even during the winter. (Most likely they're all wearing jeans nowadays.)

There was no problem with the stairs during the day, as there was a skylight window in the roof just above the top of the stairs before the U-turn was made, and its light just about reached the bottom. I can remember it was always nice and bright when the sun was shining.

Not only was our home small, but it was also mostly an *attic*. This meant that the outside wall became the roof and sloped upwards and inwards starting about two feet from the floor, toward the door, reducing the livable space of the occupants.

Our one-and-only "window" was on the north side of the roof so the sun never shone into our home! This small skylight was immediately above what we classed as our dining table. The window was about 16 inches by 24 inches (64 x 96 cm.) in the sloping wall, and could be opened and held at different heights by means of a metal bar with holes located over a peg. Everything was dark and dirty. Not much daylight entered the place as the weather was usually cloudy. Even on the rare occasions when the sun did shine in during the late afternoon and evening, it was so dismal in the room that it didn't make that much difference. I can't remember seeing anything that had ever been painted or wallpapered.

I think it is strange today to hear of people with modern skylight windows complain of how they leak. The building construction date on our attic window sill (made of *lead*, not wood) was 1745, the year before the Battle of Culloden, and it *never ever* leaked, even with all the rain we got! In addition, the walls of the building were two and a half feet (76 cm.) thick and the window frame was iron. Something else I've just thought about—the plumbing that I drank water from for the first 18 years of my life was made completely of lead. The underground pipes, the pipes up to the sink, were all solid lead—so we didn't have to worry about a minute trace of lead in the soldered joints of the modern copper plumbing! (Maybe that's why I'm half nuts!)

~ * ~

In those days it must have been very difficult for my mother to get a job, never mind hold one, as she didn't have much of an education. Besides that, few jobs were available. Mother could read and write. She used just *simple* words, but any letters she later wrote to me when I was in Canada, were always spelled properly. For example, she *never* made a mistake in the use of "two"/"too"/"to", "off"/"of", "here"/"hear", "your/"you're" or "there"/"their"—very common errors in present-day usage.

My mother's name was Christina (pronounced "Christine-ah" not "Chris-teen-ah" which you usually hear today); but she was always called "Chrissie." She was a very small woman. I don't think she even made four feet, ten inches in height (147 cm.); yet she often went down on her hands and knees to scrub other peoples' stone doorsteps (and floors), then applied some "cardinal" polish (red) and "pipe clay" which was white; around the entrances to make them look nice for passers-by. For this hour or two's work she may have received a shilling or so (equal to fifteen cents in those days); but when you have nothing, anything is a help. It was just about enough to buy a loaf of bread. She was most likely cheating the welfare system a little in doing this, as she was earning (?) money. She certainly wasn't what they call now-a-days "street smart." She was simply trying to survive in a system that made it very difficult.

I was reared almost totally on what is now called "social assistance" until I reached school-leaving age, when I was able to get a job and purchase a few things. I hated, as a wee boy of school age, being taken by the hand, quite often, to a government office which my mother and Granny called the "means test" so that I could get a pair of shoes. This office was part of the municipal offices on Witchburn Road. The means test was a system that searched into your background and checked any means you had of making money.

If you had anything of worth you were told to sell it, or if you had any income of any sort then it interfered with whatever money you got from the authorities. It should really have been called the "great degrading ordeal." (Maybe it had to be this way, for later on in life I *did* know of people who "never worked, nor wanted," who abused the system, and who were better off than I was. This was when I was married and working! These are the people who, in my opinion, are parasites in the community, and tend to spoil things for other honest people who rely on programs like social assistance to survive.)

When we got to the means test office, I had to show the man behind a desk that there was a hole right through *both* of my shoes, and, of course, my socks too, so that he could see the skin of my feet. (I didn't ever bother to count the number of times that I got a rusty nail or a sharp piece of glass into one of my feet—and getting a tetanus shot was unheard of!) As soon as one of my shoes got a hole in it, I would put cardboard inside, but had to make sure that it was removed before I went for the inspection. (I guess there was no point in going to see the man with only *one* shoe holed). If I passed inspection, my mother would get a slip of paper to take to the shoe shop for what *had* to be the cheapest shoes available.

This means test ordeal must have been very embarrassing for my mother. First, because the piece of paper, which was a government form, told the shop assistant and anyone else in the shop, that she was destitute. No doubt she had to do the same thing with my brother. There must have been other times when the ordeal was required for herself and Granny. They had to prove that they had *absolutely nothing*, and were fully dependant on the state. (It wouldn't be hard right now to shed some tears for her, for she had a hard, hard life.)

I don't know if there was a waiting time between "grants." Much later, I wondered why we had to go about for so long with shoes in such terrible condition. I know, too, that by the time I got new shoes, the ones I was wearing were too small for me and were crowding my feet. My feet still are in a "not too good" condition. Surely, I later reasoned, if there were holes in the shoes and they "qualified" to be replaced, why was it that I had to go about for ages before I was able to obtain another pair? Was it because it was only eight months since I was issued a new pair? (This figure is just for example; for I've no idea how long my shoes lasted.) I can't remember if I got a new pair of shoes every time my mother took me to that office. Maybe she was told that she would have to come back in so many weeks (or months) before she would be able to apply for further assistance.

Walking was quite strange (and difficult, too) for a couple of weeks after getting new shoes. The heels of my old shoes had been worn away to close to 45 degrees from the inside to the outside. Over a prolonged time my feet got used to walking and running at that angle. There were times, I know, that the leather on the *outside* of the upper was beginning to wear as there was no heel left. Even today, I can be standing while doing something and occasionally one (sometimes both) of my heels is over at that angle, the outside of my foot taking the weight and the inside raised off the floor, especially if I'm wearing slippers. Surprisingly, it feels quite comfortable. Sometimes, while I'm sitting at the computer, both feet are on the floor—not flat on the floor, but over at 45 degrees towards the outside. I've placed them that way because it's comfortable—like right now as I'm writing this!

The man at the means test office wasn't very friendly when he spoke. It is obvious now that my mother was a frequent visitor and that he knew her well, because he called her "Chrissie." He acted as if he were "Lord and Master" and far above us in importance. Often I'd see tears in her eyes after he gave her a hard time. (If that "Hitlerite" in the means test office was to learn of the few pennies my mother earned on the side, he would have stopped her welfare money.) I would have *loved* to have met him in the street many years later! Although maybe that wouldn't have really been a very good idea, for I likely would have been arrested for some such word or act of revenge.

In those days, people like him ruled the roost. Ordinary people were afraid to say too much in case they were "cut off" from whatever benefits they were receiving, and this was very easily done. I hope that sort of thing is not tolerated today in any form. People shouldn't allow those in office to "talk down" to them.

Understandably, the holes in my shoes were extremely uncomfortable in the wet, cold climate. So, whenever it rained—and it rained almost every day in the wintertime (and lots in the summertime too!)—my feet would quite loudly go "squish, squish, squish" as I walked. Water was sucked in through the holes and then forced back out when my foot went down again, (sometimes over the uppers, too, when it was pouring heavily and the abundance of puddles couldn't be avoided).

With my short troosers, it was obvious to anybody who saw me that my shoes had holes in them. As soon as I got into our home I would take my shoes and socks off and hold the prune-like soles of my feet close to the fire (if there *was* a fire), hang my socks somewhere close (after wringing them out) and prop my shoes next to the fire, hoping that they would be dry for school the next morning. Sometimes there wasn't any room left for my socks, as other things were drying, and sometimes there was no coal, so, no fire. Occasionally I had to put on damp or even wet shoes and socks the following day. It wasn't too bad after the socks were on, but it seems to take forever to pull on damp socks. To make things worse, I can't remember ever having more than one pair of (holey) socks!

A significant memory comes to mind at this point. My English teacher, Miss Sharpe, told me a couple of times during my school years that I should become a journalist after I finished school. (She had remarked often about the quality of my essays and compositions.) Huh, me a journalist—I who had just about no clothes on my back! What a picture that would be—*me* in an office with holes in my shoes and no underwear!

~ * ~

When I was around this age, I used to have a pronounced stutter. One day, either Ian MacKenzie or Hughie Campbell (I can't remember which) was with me around "the point" (an area just around the coast across from the north end of Davaar Island) when he got me to sit down on a rock and told me to repeat words after him. We did this for about half an hour. When we were finished, my stutter had disappeared—amazing, isn't it? I've learned in later years that many children who were "corrected" for being left-handed developed stutters. I don't remember if I was corrected at school for wanting to write with my left hand.

~ * ~

There was a gizmo that people used in Britain, called a "clothes horse" by some and "winter dykes" (a fancy name), by others— the ones that thought they were a little better than most folk. It was loaded with clothes and placed in front of the fire to dry laundry inside the home during inclement weather. This consisted of two wooden frames that had three or four cross members, joined by hinges in the middle vertically to make a self-standing unit, usually was about three and a half to four feet wide and about the same in height. The bad thing about those things was that when they were being used, no one could get near the fire! (The modern equivalent is called a "drying rack" and is a sort of concertina gizmo, with vinyl coated doweling and opening from the floor upwards.) It holds a lot more and is much more sensible, but I doubt if it would be put in front of the fire!

Other people, the ones who had kitchens, had what was called a "clothes pulley." This was a sort of indoor clothes line that was suspended from the kitchen ceiling. I'll never know why it was called a clothes *pulley*, as the only connection it had with a pulley was because the gadget was raised and lowered by means of two pulleys. It was a long length of two by two inch (5 cm. x 5 cm.) wood with a short cross piece at each end, making a very long "H", and two lengths of rope, one on each side, from the tip of one leg of the "H" to the far away leg, to hang the wet clothing on. Alas, we didn't have any of these gizmos either!

~ * ~

It's funny, as I try to recoup old memories and relive life as it was then, for the life of me I can't remember *ever* getting a new shirt, or any new clothes of any kind, when I was at school. The jacket I wore was usually either too long or too short for me, especially in the arms, as it no doubt once belonged to someone who was not the same size as I was. I can remember my mother bringing something home occasionally, handing it to me, and saying, "Here, Son, try this on." It didn't matter if it was droopy or tight, as long as I could put it on, that was all that mattered. Everything I wore was a hand-me-down from who knows who or where! In addition to never ever having any undershorts, we also didn't have any undershirts, ever. (I'm glad today that nobody handed down undershorts!!! And I still never, ever wear undershirts—even in the frigid Canadian winter.)

~ * ~

Speaking of winter, that brings up Christmas. Ah, Christmas! That time of year was a "non-event" for us. The day would come and go and I didn't know a thing about it for years! Then I found out that there were kids who would get a toy cowboy outfit that had a cowboy hat, a belt with a holster for the shiny revolver and maybe spurs. The poorer folks would share the things among the family members. Using the above for an example, one would maybe get the hat, another, the gun-belt, another would get the gun and the spurs and then they had to take turns with them!

I was about six or seven when I learned that there was a man dressed in a red suit who would come and give good children a present—and I wondered why it was that I didn't get anything as I didn't do anything bad. Gradually I learned that the man in the red suit was only a story—a farce—a great big lie. Then I didn't feel so bad. It's no wonder that I still don't have a great deal of love towards the occasion, or Santa Claus, for I know there are still lots of kids around today who get the very same as I got back then—nothing!

Today, my heart goes out (not really!) to people reminiscing about Christmases long ago and proudly stating how they didn't get very much compared to the kids of today, that they only got a little doll, or they only got a children's bake oven or something simple like that—or maybe that their parents could only afford a chicken for the Christmas dinner as a turkey would have cost too much. (I don't think I knew what a chicken was at that age and if I did, I would probably have thought it was food for a king!) If I can remember right, my first Christmas present was an orange—and that was from the Salvation Army Sunday school when I was eight or nine years of age! Yep, some people didn't know they had it so good!

~ * ~

Despite the lack of money, our fireplace grate, for some reason I've never been able to figure out, had to be "black-leaded" every week. This meant spending scarce money, when there was a little, for the *black lead*—money that should have been used for food. "Black-leading" meant smearing some stuff resembling black shoe polish on the cast iron and buffing up the ornamental bits with steel wool for contrast. This "beautifying" was something every woman of the house (my mother included) did irrespective of how ornate the grate was—or in our case how simple it was. This was meant to impress visitors, something like "look at how pretty *my* grate is!" and during my time in the town a few years later, I did see some absolutely gorgeous ones. In our case it was a total waste of money as the grate we had was just a simple place to hold coals and no one ever came to visit to see it anyway! Oh well, maybe it was a case of my mother hoping that someone might call—sometime, and that the grate had to be nice—just in case!

~ * ~

We all went hungry enough as it was, and that was *with* my mother earning a little extra on the side. I can remember quite often going out into the countryside to steal a cabbage, a turnip or some potatoes from a farmer's field, very often filling my belly with raw turnip, potatoes or cabbage before taking some home. (I always had one of our knives with me to cut off the upper growth of the turnip). I would wash a potato in a burn (Scottish for "stream") and eat it like an apple. I think I ate more raw turnip during my early life than I've ever eaten cooked turnip!

Sometimes my brother and I would go down to the pier to try to scrounge a couple of fish from the commercial fishermen when the fishing boats returned with their catches. There were so many fish in the water in those days that the fishing boats' holds were always full; so occasionally they didn't object to giving a couple away! Sometimes it would be mackerel, cod, whiting, maybe herring; quite often it was flounder (a large, flat fish) as they would be "strangers" to the catch, and would interfere with boxing the fish. The fact that the fishermen knew that we had nothing helped them to be good to us now and then. If we weren't able to scrounge any in an honest fashion, then we would steal a couple. This wasn't hard either, for the heavily-laden fish boxes lined the pier, all full of fish—some of them just *begging* me to take them home!

My brother and I would also go down to the shore and dig up cockles to take home for a meal. Sometimes, but not very often, we had to eat them raw as there was no coal to make a fire to boil them on. They weren't too bad that way, really, just a little rubbery and a little salty, and it was very difficult to open the shells. (Remember that they were still alive!) Quite often we would go to the shore to get periwinkles or whelks, but they had to be boiled before you could pick out the inside with a pin. There were always lots of mussels on the shore, too, but for some reason we didn't go for any. Maybe they weren't very edible; maybe my mother didn't like them. I don't know; but they were there by the thousands.

There were times when I would go down Kilkerran Road, sometimes with my friends, Hughie Campbell, Ian McKenzie and Ian Brodie, but more often alone, to "The Dhorlan" to chip limpets from the rocks at low tide and start a good little fire using dry sticks. When the fire had died down to glowing embers, I would put the limpets upside down on the embers and use their tiny shells as miniature pots. After ten or so my belly wasn't empty anymore; but I had to be careful, as they were quite rich, and too many could make me sick.

The secret to getting the limpets was a quick movement with a sharp-edged stone. If you happened to touch the shell, even ever so lightly, just before the sharp flick, then the limpet got a really good grip. You would actually see it sucking itself tight to the rock. In cases like that you forgot about that particular one as there were lots more around. If you were silly enough to continue in your quest for that particular one, you would smash its shell to bits and it still wouldn't let go. (This is where the limpet mine that commandos attached to enemy ships during World War II got its name.)

* ~ * ~ *

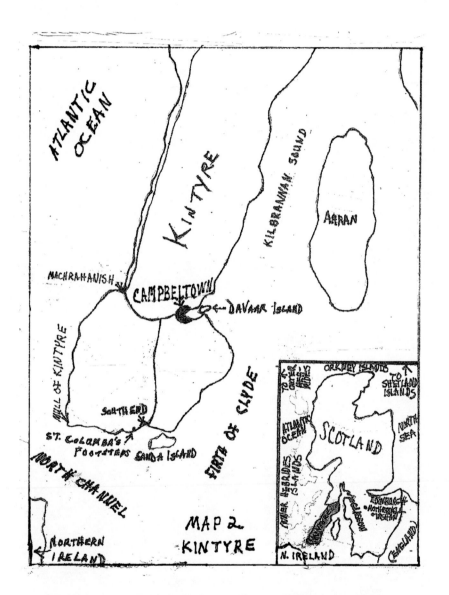

Kintyre peninsula with insert of Scotland, highlighting Kintyre

CHAPTER TWO

During and Just After the Big War

In school I learned the history of our area and found it fascinating. Embarrassingly shaped like an enlarged penis (sorry, but that's the truth; and, no, I didn't learn that in school!), Kintyre is the southern-most peninsula on the west coast of Scotland, very close to Northern Ireland. The Mull of Kintyre is the headland at the south west tip of the peninsula. My hometown, Campbeltown is on the east coast of the peninsula just northeast of the Mull of Kintyre. There is the possibility that a village of some sort was at the location of present Campbeltown at the same time that Christ was walking this earth! Originally it was known as Kilkerran, the capital of that area when it was peopled by Celts (the original Scots). Northern Ireland is just a short boat ride from the Mull of Kintyre (13 miles /20 km) Later, while still a village, the name was changed to Dalriada (pronounced "dal-ree-ah-dah") around 500-600 A.D. The town became a "Royal Burgh" around 1780 A.D. when it was visited by members of the Royal Family.

Kintyre is the area where St. Columba (521-597 A.D.) landed prior to building the abbey at Iona (one of the Inner Hebrides Islands off the west coast of Scotland). He was partly responsible for bringing Christianity to Scotland. The Ionian monks were slaughtered by the Vikings and the abbey ruins are still there. There are some steps at Southend (a village on the tip of the Mull of Kintyre, ten miles or 16 km. from Campbeltown) that has St. Columba's footprints etched into the stone. If you are ever in the area, just look for signposts that read, "To St. Columba's footsteps."

You see, St. Patrick went from Scotland to Ireland and St. Columba from Ireland to Scotland. They both became missionaries in their adopted countries. (Maybe they were on a sort of "lend-lease" program! Either that or St. Paddy was fishing in Campbeltown Loch and a strong wind arose and blew him over there so Ireland felt that they had to go tit for tat! St. Patrick was a boy when he was kidnapped from Scotland by Irish raiders.)

The Mull of Kintyre is where some Viking chief made a crafty deal away back around 700 or 800 A.D. (probably the same bunch of Vikings that slaughtered the monks). It seems that an agreement was made with the King of Scots for peace with the Vikings. The "Horned One" was to be given all of the land he could get his ship around and the agreement was that he would then leave everyone else alone. Well, the crafty devil sailed around *all* of the Hebridean Islands, then southwards down the west coast of the Kintyre peninsula, right around the Mull of Kintyre, and then sailed north up close to where the village of Tarbert now stands on the northeast side. He then had his men *pull* his boat over the half mile of land that prevents Kintyre from being an island! So Kintyre was added to his quota!

~ * ~

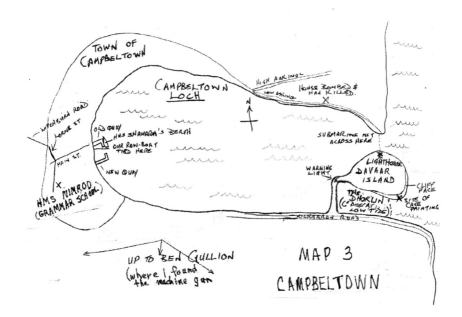

Campbeltown Loch, a sea-loch

The loch at Campbeltown (a sea-loch) is in the shape of a horseshoe lying on its side with the opening facing east. ("Loch" is the Scottish word for "lake." It has a guttural "ch" sound similar to that in the German word "ach." The sound is made by forcing air between the back of the tongue and the soft palate at the roof of the mouth. This is a totally alien sound to most English-speaking people, who generally manage to say "lock".)

Campbeltown Loch is about three kilometers (two miles) in length from the opening of the "horseshoe" to the harbour at the western end. Guarding the mouth of this haven for sailors during rough weather is Davaar Island (two syllables, Da-vaar—emphasis on the "v").

The first lines of a song about Campbeltown Loch, I would like to think "us wee boys" in Campbeltown gave to a certain wee man as the idea for a hit song he made famous in the 1960s— "Campbeltown Loch, I Wish Ye Were Whisky." (Some claim it was an old Scots folk song or a song based on an old pipe tune; others that it was written by Andy Stewart.) Whatever the truth, renowned Scottish entertainer Andy Stewart (now deceased) made it a very popular song in Scotland, possibly all over the world. You see, as we were growing up, three or four of us would go arm in arm down the street singing the first few words—"Campbeltown Loch, I wish ye were whisky"—that's all we knew at the time. I like to think that Andy heard those few words sometime in Campbeltown and created a song around them. *"Oh, Campbeltown Loch, I wish ye were whisky, Campbeltown Loch, och aye! Campbeltown Loch I wish ye were whisky, I would drink ye dry!"*[1]

The verses cleverly have the singer imagining how nice it would be if the loch were full up to the brim of whisky and he could anchor a yacht in the whisky-filled bay to go in for a nip and a dip "by night and by day." Clan gatherings would feature wading into the loch with toasts of "Slainte Bha" (pronounced "Slanj-eh-vah"—good health). The only problem would be the police showing up in a boat and shouting, "Time, Gentlemen, please!" I find this a fitting tongue-in-cheek ode to a town that once boasted of 30 distilleries and still produces at least two very fine brands of single malt whisky – Springbank and Glen Scotia.

(I'm going to jump many years ahead now to the time that my wife and I went to hear him when Andy Stewart was performing in Winnipeg, Manitoba. When he was exiting the theatre, I went up to him and asked if I could shake his hand. That got his attention! I thanked him and told him that he had put my little town on the world map. Then I told him the story of us boys singing the only bit that existed away back then. Now back to the history lesson!)

~ * ~

---

[1] Chorus of "Campbeltown Loch, I Wish Ye Were Whisky" a Scottish folk song popularized in the 1960s by Andy Stewart (1933-1993).

Davaar Island, where the cave painting of the Crucifixion is located. The cave is at the upper right hand of the picture. In the foreground is a meadow at the edge of Campbeltown Loch along Kilkarren Road.

This rocky island (Davaar) is just a little less than five kilometers (three miles) in circumference, and there is a beautiful lighthouse on the north face to guide ships to the correct side for safe entrance to the harbour. The east and southeast sides are made up entirely of cliffs. The gap between the south side of the island and the mainland will not allow anything other than very small boats to pass into the loch, and that is *only* at high tide.

There is a very interesting tourist attraction on the island which wasn't found for many, many years. Davaar Island has quite a few large caves, especially on the south east shore. In one of the more remote larger caves a local artist by the name of Alexander McKinnon did a life-sized oil painting of Christ on the cross on a large, flat, vertical rocky face, well inside the cave from the entrance. This painting, done in 1887, is very realistic and absolutely beautiful. It wasn't discovered for many years, partly because of the remoteness of the cave, plus the fact that Mr. McKinnon kept his creation a secret. McKinnon must have been busy there for endless hours painting it by lamplight. He'd have to have had a supply of paint, food, clothing, drinking water, sleeping blankets (nothing akin to sleeping bags in those days) and other necessities (toiletries, for example) to allow him to remain inside the cave for who-knows-how-long to continue his work! Complicating it all, the tide ebbed and flowed twice every 24 hours and 50 minutes (a *lunar day*). Also, he must have walked back and forth from the town with his requirements, taking the "path" over to the mainland at low tide to replenish his stores. If he had taken a rowboat to save himself the walk and inconvenience of carrying his goods, some of the local fishermen would surely have seen the boat up on the shore and would have investigated, thinking that someone was having difficulties. Thereby they would have discovered his activities. However, he seemed to prefer to make his contribution to society in secret.

Many years later a request was put out in the media for the artist to return, whoever he was, to touch up the painting, as it was starting to deteriorate badly because of the damp, salty weather. The artist returned in 1934, made himself known, and totally renewed it. I don't know if it has been retouched by anyone in recent years, but the answer should surely be a "yes", as it is a must-see for tourists (if they don't mind clambering over lots of rocks).

~ * ~

Davaar Island is joined to the south shore at low tide by an upside down "L" shaped sandbank. There is a beacon to warn ships (a miniature lighthouse) on the part of the "L" that changes direction.

Using this sandbank as an access footpath to the island allows you to do a quick visit, maybe have a nice little picnic on the grassy slopes that meet you as you end your trek over this "path." Doing so means that careful note has to be taken of the time of the tides. You need to make your way there as the tide ebbs and bares enough sand to walk on (about a quarter of a mile to the mainland). The return journey needs to be made back to the mainland in plenty of time, otherwise the unwary will be stranded until the next low tide! That did happen quite a few times during my youth, and always to visitors to the town who didn't know just how fast a tide can rise. It almost happened to me one of the times we were a wee bit tardy in leaving. My companions and I had to wade through about three inches of water before we got to the mainland.

Sometimes the "stranded" would be spotted by passing fishing boats returning to town. They would then tell Jimmy O'Hara who, in turn, would rescue them with his motorboat. I don't know what he charged for this service, but I do know he wouldn't do it for free. He had a little sideline taking people out to the island and dropping them off at the small jetty that serviced the lighthouse, but only during the tourist season. He would ferry passengers from the town to the lighthouse and pick them up at a later, prearranged time. This had no relation to the tides. If the stranded people could make their way round to the lighthouse, then it was better for them, as they then had a chance of being spotted and rescued. It was too bad for them if the last "pick-up" had been made. They then had to hope they'd be spotted by a passing commercial fishing boat. Occasionally an S.O.S. would be sent out, as friends knew their destination, and a special trip would have to be put on to rescue them.

The area just east of the sandbank (called "The Dhorlin") and south of the island was very shallow at low tide. As boys, my friends and I would very often take our "spears" (a long nail firmly bound to the end of an old broom handle, or a reasonably straight branch), remove our shoes and socks and wade into the 6-8 inch (13-20 cm.) deep water very gently. Looking carefully, I might spot a flounder lying quietly on the bottom (very well camouflaged). I'd trap it with my foot and push the "spear" into it. Then I'd take it out of the water after putting a hand under it while it was still on the bottom so that it wouldn't slip off. A couple more and I had enough for dinner. An odd time the "spearist" was a bit too keen and would end up spearing his foot! Ouch!

~ * ~

The Second World War started when I was seven years of age, but it barely touched our town, which had quite a large naval base. How the Germans didn't know of it and bomb it repeatedly, I don't know, for they had spies everywhere.

Very often the loch was loaded with warships such as destroyers, minesweepers and submarines. There was a submarine net strung across the entrance, from the island to the north shore, to prevent U-boats from going in and picking off the ships. The loch was beautifully situated for easy access to the Atlantic Ocean.

During the war, the Grammar School building had been taken over by the Admiralty as a training centre and had been renamed "HMS Nimrod." Because of this, our classes were held in what are called "portables" today.

During the war in Europe, Germany had a radio propagandist who we Britishers called "Lord Haw-Haw." He would broadcast to the British people that Germany was going to bomb specific areas of the country or boast of targets that had been hit. I believe he was a traitor Englishman and I'm struggling to remember his proper name, William something. He knew many landmarks, mentioning in the broadcast by name which one was going to get bombed *that night*—anything to try to demoralize the British. One night, maybe about half way through the war, this character broadcast that German U-boats had just sunk the British cruiser HMS Nimrod, which, of course, was (our) Campbeltown Grammar School! (Our family almost missed hearing the broadcast as we hadn't owned a radio just prior to this and had just been given an old battery-powered one.) We all thought the announcement was hilarious! I was told it was put out in the papers across the length and breadth of the country that the great German Navy, according to Lord Haw-Haw, had managed to sink a school that was in the middle of a town! This news actually had a positive bent, as it boosted the morale of the British people, letting them know that a lot of Lord Haw-Haw's broadcasts were just that—"haw-haw!"

I can remember the uproar in my town when Italy joined forces with Germany. Campbeltown boasted three Italian restaurants. After this news, these restaurants had their windows broken, doors burst open and interiors destroyed by a crazed mob. The people who owned the shops were terrified for their lives, and they were really very nice people. It's amazing just what a few hate-mongers can do!

~ * ~

During the war, workmen went all over town removing anything made of iron that could be cut and taken away to make guns, tanks and anything of steel. Any church, house or establishment that had wrought iron railings and gates was left bare (unless removal caused a safety hazard). Very little wrought iron railing was left anywhere. If a person went to the town dump, he wouldn't find one bit of steel anywhere. It was all needed to fight the enemy.

One evening, early in the war (I was probably about eight years old), my grandmother and I were sitting at home, just the two of us. Since it was dark outside, it must have been wintertime, as it was still quite early (maybe shortly after six o'clock). We were having a meal of fried sliced sausages. I can still picture the *two* slices of sausage on my plate—nothing else mind you—but it must have been dinnertime, for *all of that* was my dinner! Suddenly we heard the sound of an aircraft flying overhead, the heavy engines droning loudly. Granny said to me, "What kind o' plane is that, Son?"

She knew that I paid a lot of attention to what types of aircraft were at the local small aerodrome (airfield) at

Machrahanish (pronounced "Ma*ch*-ra-han-ish"—same guttural "ch" sound) as it was a training school for the Royal Air Force. I listened to the droning sound for a moment and told her, "Ah don't think it's one of o' ours, Granny."

Well, the words were hardly out of my mouth when we heard "boom, boom, boom"—just like that. It was the first three of the nine or ten bombs that fell on Campbeltown during the entire war. And this was only because our town was on the flight path from Germany to the shipyards in Belfast, Northern Ireland. If there were any bombs left in their bomb-bays on the return journey, they were dropped wherever a light was seen. One of the bombs hit the Royal Hotel, taking away the top of the south corner and the other two landed in the harbour and did no harm. The siren didn't sound that night so there most likely were lots of lights showing. A couple of times the streets were strafed with machine gun fire during the night and that was it. The pilots probably saw a light, swooped down, fired, and then continued on their way home. Only one man was killed when his house was flattened by some sort of land mine at a later date, and his home was about two kilometers from the town.

I mention "seeing a light" because there was a complete blackout while the war in Europe was going on and it was imperative that no lights were made visible to enemy aircraft. Everyone, all over Britain, had the responsibility to make sure that the blinds and curtains were tightly drawn before the inside lights were turned on. Air Raid Wardens used to patrol the streets after dark looking for houses that were showing even a "chink" of light. I can remember how wonderful it was when Germany was no longer a threat. The street lights were turned on for the first time in years, and the windows of the houses were allowed to help light the sidewalks at night if the curtains were left open.

I've been trying to remember how it was that our window was "blacked out" but I just can't think how. I know for sure that we didn't have any curtains, as they would have just hung straight down from the skylight, but there must have been some way to stop the light escaping. Maybe we just blew out the oil lamp or the candle when the air raid siren went off, which was most likely, and sat in the dark until the "all clear" sounded. This happened quite often, maybe three or four times a month.

The McMillan family (whose son George became a good friend of mine) moved out of town to live in Tayinloan, a small village 17 miles north to avoid the possibility of danger because of the nearness of the Naval Base, fearing that the town would get bombed a lot. It was sadly ironic that the second oldest son, John, was killed by a vehicle a few weeks after they moved there. It was even sadder because the father was a prisoner-of-war at the time and I doubt if he was aware for a long time that he had lost his son, perhaps not until he returned home when the war ended. Maybe prisoners-of-war received mail, I don't know for sure. I think the father must have been one of the first British soldiers to be captured, for I know that he spent almost the entire war as a prisoner. It was like he put on his uniform in 1939, walked into Germany and yelled, "Here I am, Jerry."

~ * ~

Watching the Campbeltown Pipe Band playing on Main Street in Campbeltown. It might have been like this in 1945 to celebrate the end of World War II.

I was nine years old when my grandmother died. She died at home in our only bed, and her body remained there for three days. Finally, some men came, put her in a coffin and took her away to be buried. I can't remember her burial. I guess I wasn't at the funeral, if there was one. We had to sleep on the floor while the corpse was on the bed. I assume Mother had borrowed a couple of blankets for us, as we didn't have any extras.

We had more room after they took Granny away. There were only three of us then. The blankets were not washed, nor were there any facilities for sterilization or anything like that. It was just a case of going into bed that night as though nothing had happened.

I don't know if Granny ever contributed anything toward the home—probably not. She, most likely, was supported by the "means test" as we were. My main memories of her were when she would come home drunk, which really wasn't often—maybe five or six times in my memory. Where she got the money for alcohol was a mystery to me; maybe it was money from the "means test" with which to buy something specific. Maybe she had only one beer on an empty stomach, which is probably closer to the truth. British beer was a lot stronger than North American beer at that time; I think around six and a half percent *by weight* not by volume. My mother told me that Granny had a job carrying beer to workmen when she was a young woman. She hauled large pails, two at a time, using a yoke across her shoulders. Mother also mentioned that Granny had the odd sip of beer from time to time, and that is how she developed a taste for it.

She would always sing an old Robbie Burns song, "Ca' the Yowes Tae the Knowes" every time she came home like that. *"Ca' the yowes* tae the knowes*, Ca' them where the heather grows, Ca' them where the burnie* rowes*, My bonnie dearie."*[2] *Scottish words: "Ca'" (call or drive); "yowes" (ewes); "tae" (to); "knowes" (knolls or hillocks); "burnie" (little brook); "rowes" (rolls).

Though I said, "came home"—that's wrong. We would be told by "whomever" that Granny was sitting, propped up against the wall of McConnachie's Garage in Burnside Street (her favourite spot). Archie and I would have to go get her and help her home, and she would be singing her song all the way. It's an old Scottish song, still available in recordings today. Every time I play it I think of her being sloshed and my brother and me helping her home and up the stairs. When I got older I didn't blame her for her wee drink. It was most likely the only escape she had from reality; for I don't doubt that she also had a very hard life, even long before I came on the scene.

Another memory is of entering a local shop and being surprised by encountering Granny there, speaking Gaelic with a shopkeeper. I had had no idea that she spoke the Gaelic and

---

2 First verse of "Ca' the Yowes," a traditional Scottish song by Robert Burns (1759-1796).

sincerely wished that she would have taught me to speak it. I don't know if she ever taught it to my mother either. Surely Granny must have spoken the Gaelic as a child.

~ * ~

There would come a time, shortly after the start of summer school holidays, when the local farmers would hire boys to thin out the cabbages, turnips and lettuces that had just started growing to 3 inches (8 cm.) or so high, in the fields. The farmers would come into town with their lorries (trucks) to pick up boys who wanted to work for them. Pick-up time was between 7:00 and 7:30 in the morning. If you were a little late in getting out of bed then everyone was gone when you got to the pick-up place, as it was important to get as much daylight as possible. The man in charge of the boys who went "thinning" was a person by the strange nickname of "Henry Pea Soup." We didn't call him that to his face, of course—just Henry—and I've absolutely no idea why he was called that. He must have been paid by the farmers for he didn't get anything from us.

When we arrived at the fields, all of us boys wrapped burlap (the same stuff that gunnysacks are made from; only we called it "potato bagging") around our *bare* knees (remember we wore short trousers), tied each one with string, then got down on our knees and got to work. We crawled along the rows at a snail's pace while we did the "thinning," which was removing excess growth and leaving only the strongest plant every so far. For every row we did we got paid ninepence (eleven cents). These rows were so long that it seemed to take forever to reach the far end of the field at the finish of each row. If a row happened to be a bit shorter, then the rate was only sixpence a row. And as soon as you finished one row, you then started to come back the other way on another row. One really good thing about our job was that the farmer's wife would arrive in the middle of the day with lots of food for us. After a good "tuck-in" and a wee rest, it was back to work until about four o'clock. Then she would appear again with more food, so we could work until an hour before dark. Then it was PAY TIME! Of course, this was the best part for most boys; but I'm not so sure about me, for I certainly enjoyed the nice food. It was *far better* than anything my mother ever made, for everything we ate was cooked over an open fire. I think there's a possibility I would have worked just for the food—really! There never was any problem in getting to sleep after a day of eleven or so hours crawling along on your knees, doing that sort of work.

The farmer kept a tally of who did what, so some got more than others. He also inspected our work and if it wasn't thinned properly he would soon let us know. If he had to tell you a second time, then he wouldn't allow you back on his farm. We also did similar work at the end of summer, picking potatoes. This time we didn't have to go down on our knees. Our job was to continually bend and fill our baskets, emptying them into a central container.

When the rows were quite long, but maybe not quite ninepence worth, a mean old farmer would insist on sixpence a row, but a decent farmer would pay a penny or maybe two pennies more than the rate for a short row. So, when the farmers arrived to pick up their help in the morning, all the boys would try for the better-paying farmers first and ignore the tightwads until the last minute. The tightwads sometimes were left with the boys that the other farmers wouldn't take, the ones that didn't "thin" properly. Thus, those "tightwad" farmers were their own worst enemies!

At that time there were 240 pennies to the British pound and I think there were five Canadian dollars to the pound. (The modern day British pound has one hundred pence.)

The work only lasted for about three weeks each time. This was a sort of measure of which families in town were in need of a little extra money, as it certainly wasn't easily earned. The work was next door to slave labour; but even then, some of the skinflint farmers underpaid or short-changed us. All the boys who did this work were around 11 to 13 years of age, and, there *never* were any girls. Seldom were there any 14-year-old boys, as at that age they were usually working full time. It was easier for a 14-year-old boy than it was for a man to find work, as they earned less. Something else I didn't realize at the time was that there were not as many men available for this type of work, for quite a few men folk didn't return from the war.

~ * ~

We had an occasional visit from a relative of my grandmother who came from Inverness. His name was Neil Killin, the same surname as my grandmother's. I didn't particularly like him, as he seemed a little "different." I realize *now* why he seemed different. One night as he was leaving our home (I must have been about eight or nine years of age, for my grandmother was still alive), he insisted that I show him "down the dark stairway, to light the matches." Half way down he wanted me to give him a kiss. Hey, I wouldn't have kissed a girl at that age, let alone a man!!! I ran back up the stairway, yelling for my mother. We never ever saw him again. Years later, I wondered just how many young boys could have been swayed in the wrong direction by such an individual.

~ * ~

Every once in awhile, especially when Mother was in dire straits—even more so than usual—she would say to me, "Go tae Jock Campbel and ask him tae lend yer mother ten shillings." (This was about a dollar and a half, but with a lot more buying power than nowadays.) But before I would go she always cautioned me to wait until he was on his own. There never was any hesitation from him. Out would come his wallet and a ten-shilling note would be handed to me. I never thought much about it until many years later. As far as I know, my brother was never sent on a similar mission and I never thought to discuss it with him or even ask Mother why. ("Jock

Campbel" was not his real name, and the use of this pseudonym will be explained in "Came to Canada, Eh?," the sequel to this story. He was an upstanding member of the community and a married man with children.)

~ * ~

I don't have many memories of playing at #8 Lorne Street. There was the "linoleum man" I made when I was about 12, shortly after a rare visit to the movies to see *The Wizard of Oz*. I made a copy of the Tin Man. Someone had thrown out some chunks of old linoleum. I collected the bits and took them up to our place. Using scissors and string, I proceeded to put together feet, legs, arms, a body, and a head, all made from rolled-up linoleum. Archie came home, and told me, "Yer affa daft"—meaning I was very stupid. I thought it looked quite good when it was finished. The arms were jointed and so were the knees, and they were the same length when I held *him* up so that *he* was "standing." That was a remarkable feat right there!

My brother always ran me down no matter what I did, so his remark didn't faze me. He always wanted to fight with me, and never once during those boyhood years did he ever tell me that he liked me, always the opposite. Of course, it went both ways; for I didn't like him either. Well, it stands to reason, doesn't it? How can you like a person who always wants to fight with you? Once, a few years later, he tried to mess up my face with a broken tumbler, and I only prevented it by kneeing him in the crotch. (It wasn't a very enjoyable boyhood!)

Other boys had brothers and sisters whom they joked with, laughed and played with, but I couldn't do that with my brother. Most times he acted as if I didn't exist, and that was okay with me. It meant he was leaving me alone. He surprised me, though, when some older boys were picking on me once when I was about 12. Archie told them to leave me alone. I couldn't believe it! (Fortunately, it seems that everyone gets a little more sensible when they get older. My brother and I became a lot closer over the years after I emigrated, even to the extent that we phoned each other off and on and when we were visiting Scotland, my wife, Mary, and I would visit his house in England for a few days where we had a pleasant time together. He was fortunate enough to be married to a very caring woman (who happens to be named "Mary" also; so there were *two* Mary Morranses, married to brothers!)

Maybe I'm overstating the situation. You see, the word "love" was not used in our normal world. My mother never ever told me that she loved me. Even when I would phone her from over the Atlantic many, many years later, on saying our goodbyes, I would say, "I love you, Mother." Then she would answer, "Aye, okay, Son—cheerio!" (British for "goodbye"). I don't think that any of us expected her to say differently; that's just the way it was.

The nearest that she ever got to doing so was during my visit to Scotland in 1976. She wasn't at home when my late wife and I arrived there as she had no idea what time we would appear, but eventually we met her in the street. (This was easy in such a small town as ours as I just had to ask anyone, "Have ye seen Wee Chrissie?") When I hugged her (I think for the first time in my life), she said, "Ah, my favourite son." I think she just said that to make me feel good, for she certainly hadn't treated my brother any less than she had treated me. (This situation was good for me, for I certainly made sure that my own family would never have any doubts that they were loved by me—ever.)

It is obvious that she cared very much for both of us, for her to work at the degrading odd jobs she did to get a wee bit extra. It's just that my mother, Granny and I didn't show any affection or caring toward each other at that time. Perhaps we didn't know how to.

The closest I came to publically honoring my mother was during that same visit to Scotland when she had greeted me as her favourite son. We were in a local pub and I got up on stage to sing "Old Scotch Mother Mine" to her. It seemed appropriate since the words are those of a Scottish emigrant, which I was. *"Though we're far apart, for the sake of auld lang syne*, God bless and keep you, old Scotch Mother Mine!"*[3] *auld lang syne" (days long ago)

I'm sure she enjoyed my singing, though she never expressed it. I sang another song after that so, if she'd had any tears in her eyes (as I did), they were gone by the time I took my seat.

~ * ~

I suppose remembrances of my boyhood are similar to those of my service in the Royal Air Force. You mainly remember the good times and try to forget the bad, and there were very few good times at #8 Lorne Street. I think of the many hard times my mother had and find it quite difficult to keep the tears from welling up in my eyes. I remember, too, that she was always humming a tune just under her breath no matter what the situation. Her life must have been one continuous worry from getting up in the morning to going to bed at night, trying to look after us and feed us. I'm proud to say that my habit of singing or whistling a tune while I work probably came from her.

It's difficult to explain that sometimes it was just impossible to get food when there were no means of obtaining any. If there were two slices of bread left, then one was for my brother and the other was for me. Her usual remark, "No, it's all right; ye go ahead and take it, Lads. Ah'm no' very hungry." just about says it all! She made tremendous sacrifices, working for "buttons" when she wasn't fit, and worrying how she was going to raise two young boys on her own in a system that treated her like dirt. When I look back now, I think of her as a queen. A destitute—uneducated one maybe—but a queen nevertheless!

In our home there certainly was a shortage of food and other comforts, but there was never a shortage of instruction

---

3   Excerpt from "Old Scotch Mother Mine," words and music by Ray McKay and Joseph Maxwell.

on honesty, manners, showing respect for others, and how to behave in society. Even without the extra education I gave myself, I could have been put in any situation and not have embarrassed her. For instance, considering that I am left-handed, "correctness" meant that I had to use the correct hand for the correct piece of cutlery—the fork in the left hand and the knife in the right hand. (Left-handers initially like it the other way around.) When I was growing up, if I had used cutlery like a lot of North Americans do, Mother surely would have corrected me! Some Canadians (and most Americans) will cut with the knife in the right hand, put the knife down, and then transfer the fork to the right hand to take it to their mouths. I find this unbelievable! Is the left hand incapable of moving food towards the mouth? Others I've seen don't even do that, they don't know what a knife is for, using only a fork as a common tool to do the cutting and the shoveling, using the fork upside down. I find it quite amusing seeing them do this in a respectable restaurant. And *I* was the one who could be considered "dragged up—not brought up!"

~ * ~

I have regrets regarding how I often didn't help at home when I was a child. My major regret was the time when I was around ten or eleven and wanted a paper cutout, three dimensional model of a Sunderland Flying Boat that I saw in a shop window. I wanted that thing and it didn't matter how many times my mother told me, "We've nae money an' ye ken fine (you know well) we canna' afford it." I wanted it, stomped my feet, pouted and cried, cried, cried for this model until I finally got it. I don't know to this day how on earth Mother managed it, but it happened. I get a feeling like a knife going into my gut every time I think of it, and it has been this way all of my life since then. That was *food* or *necessities* money! What she had to do to accomplish this, or how she had to sacrifice, I didn't even consider at the time. I can remember making the model and gluing all the tabs (a tube of glue was included in the box). It did look very nice when I had it finished, but I can't remember what eventually happened to it. That shows just how important it was. It was a couple of years before my conscience really started to bother me regarding the paper model airplane, probably because I was still quite young and didn't fully understand that we had *no* money at all and that it was *extremely* difficult for us to get any. My selfishness was inexcusable and I have regretted it ever since.

I like to think that I made up for it in a way because, when I was a little older, I found a way to make a bit of money. I'd locate the better class homes that had just taken a delivery of coal, go to the door and ask if I could shovel it into their coal cellar. Usually it was the "better off" people who paid the least for getting this job done—those known then as the "gentry folk." It seemed to me that the poorer the people were that you were dealing with, the more generous they were. For doing this heavy shoveling, I earned according to how much was delivered.

For a quarter ton (560 pounds) the payment was usually a shilling, sometimes just ninepence (what the gentry folk paid, but they usually got a ton delivered anyway). For a half ton it was one shilling and sixpence, and a full ton warranted a half crown (two shillings and sixpence). It was 12 pennies to a shilling, 20 shillings (or 240 pennies) to the pound. So, if your math is okay, you'll see that it was always nice to get the quarter-ton jobs, as four, equal to one ton, netted four shillings. (A "ton", *imperial*, commonly known as the "long ton", is the British measurement and is 2,240 pounds. The North American ton, the "short ton", is 2000 pounds and the metric *tonne* is 2,204 pounds).

The thing was that I couldn't wander all around town looking for the quarter-ton better-paying jobs. It wasn't every day that people got coal delivered, so I grabbed the chance whenever I saw a pile of coal at someone's house. If I waited until later, then someone else would be shoveling it when I returned. Some days, especially during the summer months when I was free from school, or on a Saturday, I could get two, sometimes three lots to shovel and go home with maybe four or five shillings in my pocket to give to my mother. There was the odd time when I had finished shoveling, the owner (*always* one of the "gentry folk") would say, "Oh, Ah hav'nae ony change, Son. Come back tomorrow and Ah'll pay ye then."

The next day I'd hear the same thing, *and* the next day, *and* the next. There they were, the elite people of the town who could afford to get a ton of coal delivered, and after getting a 12- or 13-year-old boy to move it for them, they were too blinking miserable to pay for the work. It's no wonder they had money. But, the old saying holds true, "once bitten, twice shy." That house never saw me again, even if I saw a pile of coal at the door. I wonder just how many young boys these people robbed by not paying them. The sad point is that if I accused those people of being thieves, they would be horrified; but that is exactly what they were. I need to emphasize that it wasn't just a matter of moving the shovel to deposit the coal in their cellar. Oh no, usually it had to be wheel-barrowed from the street first. It was *very* hard work!

The bad thing about doing this shovelling was that I would get so dirty (except when it was raining, then there was no dust). I often did this kind of work in the rain as well as when it was dry, getting soaking wet at the same time, as though it wasn't hard enough work! The more coal that I shoveled, the dirtier I got, so that I soon looked like I had been down in the coal mine digging out the coal myself! And it all had to be removed with *cold* water. Brrr!

To make it worse, the soap in those days wasn't very good and it certainly wouldn't lather up in the cold water. It was pretty close to, but not quite, the equivalent of lye soap, which was used to wash floors and clothes and the like. I have a recollection that it might have been called "fairy soap"! Boy, could I make a humdinger of a remark on *that* one! Maybe call it "Cousin Neil Killin soap"!!!

When I went outside afterwards to play with my pals, they would know immediately that I had been "doing the coal," as, try as I may, there was always the telltale dark around the eyes and in the ears. It took a while to get all the fine black dust out, and then it was back again with the next shoveling. Not that it was anything to be ashamed of, for lots of boys went "doing the coal" to get extra money, and they were usually identified right away. There was no one else I knew of who was chided by the others, "When ur ye gan tae steal any mair o' the fairmer's tatties, Ian?" (Translation: "When are you going to steal any more of the farmer's potatoes, Ian?") Nobody else I knew ever got that thrown at them; so my brother and I must have been the only ones in town who did that. Of course, those remarks weren't half as bad as the occasional pronouncement, "Yer mother's a whore!" from some of my pals. They had, no doubt, come by this "information" from overhearing some of their parents' gossip.

I can remember being taken out in front of the class at Millknowe Public School when I was twelve, by my teacher, Mr. Gemmel, for the rest of the class to see just how dirty I was. "Look, boys and girls, look at this dirty boy!" His educated "gentry" talk held me up to ridicule and embarrassed me by parading me for all to see and laugh at. Then I was told to go and thoroughly wash my face at the hand-basins in the school hall, using their soap and hot water.

Funny though, I thought it was great! I took my time and had a good wash. At a later date I tried the same thing again but either the teacher had caught on to my idea or he figured that there was too much time lost. There was also the possibly that a stink was raised about the coal-black sink I left for someone else to clean! (Cleaning the sink after I had dirtied it never entered my mind until many years later.) I can't really be blamed, for I was used to an old black cast iron sink that didn't show the coal dust, not a nice white porcelain one. And, no doubt, the nice white towel on the roller wouldn't have been white anymore!

One Saturday, I was very lucky and got three lots of coal to shovel (this amounted to 6720 pounds or 3000 kilograms and a *lot* of work). I went home to give the money to my mother and she wasn't home. So, instead of taking the three half crowns outside with me, (seven shillings and sixpence, about a dollar and ten cents, and this was really a lot of money for us), I hid them in the drawer and went out to my pals. Later in the evening when I was at home I went to get the money to give to my mother and it was gone. I never saw it again. Bill Moorhead was living with us by this time and I don't know whether it was he or my brother who had taken it, so I can't very well point a finger at anyone. Neither of them would own up to taking it. (Oh well, that was another lesson learned.)

When my brother finished his schooling and started work at 14 years of age, he gave Mother some but not all of the money he earned. His earnings certainly helped, and it was even better when I started work, as I gave Mother *all* of my wages and she gave me some pocket-money back. (It's *amazing* what the shameful memory of a paper model airplane can do!) I shouldn't be too hasty in judging Archie, for he does deserve some credit. Later on, after I had emigrated, he was there when he was *really* needed and I wasn't able to do my share.

~ * ~

I think I was about twelve when the following happened. Just to the south of the town, and bordering on it, is Ben Ghuilean (the Gaelic spelling; normally now it is referred to as Ben Gullion. The word "Ben" in Scot's English means "mountain.") This is a reasonably-sized mountain. I have already referred to a small airfield five miles from town. This airfield was still used after the war to some extent for training Royal Air Force pilots.

One foggy day a two-seater aircraft plunged into the side of that mountain, killing both airmen.

It was quite a climb to the crash site and, needless to say, there were lots of (morbid-minded) townsfolk who just *had* to make the climb, though they would never have considered doing so at any other time. Apart from the strenuous effort, it was well known that there were adders on the mountain. (Adders are a type of viper, a little over two feet long. The bite of this snake, while it wouldn't kill you, would make you very ill for some time.) This thought didn't bother us brave (or stupid) lads, as we spent quite a lot of time on various faces of the mountain. (I had killed an adder some time before and preserved it in alcohol in a glass jar to keep in the house. No one objected at first, but later I had to keep it where we kept our coal.)

No one was allowed anywhere near the crash site until the bodies of the two airmen were removed. People were collecting bits of this and bits of that—stuff that probably went into the rubbish bin (garbage) a few weeks down the road after they had lost interest in the incident. Not so with "yours truly." I noticed that there were two machine guns, one on each wing, and I set about removing one. What did I want a machine gun for? Maybe I was going to take it to class for "show and tell." Na, we didn't have that silly exercise in those days. I really had no idea why I was taking it. I guess it is what the modern kids would call "cool."

Anyway, I struggled with it for ages and finally got it free. Even today, I still marvel at the fact that I got a machine gun from an aircraft without having a spanner (wrench) or even a pry-bar. I carried the heavy thing down the mountainside on my shoulder to the foothills, where I hid it by throwing it into the middle of some "whin bushes" (furze or gorse). These bushes were evergreen, covered all over with long, sharp dark green needles, standing three or four feet high and at least that across, with nice yellow flowers. (They grow wild in Scotland, but I don't believe they grow in North America, unless maybe on the east coast.)

I hid the gun because it was still daylight and I didn't want anyone to see me walking into town with a machine gun over my shoulder. Besides, I had to walk past the police station! I would probably have been arrested (or worse still, maybe even *talked* about). So, what did I do when it was time to retrieve it? Well, I got hold of some old potato sacks (gunnysacks), my friend Ian McKenzie and his four-wheeled cart, and the two of us headed back up to where I had hidden the gun.

What do you know? *It wasn't there*! Did I have the correct bush? "Look over there …. No … try this one … ." There were lots of clumps of bushes. We just about went crazy! I was quite sure that I had taken note of where I had hidden it so that I would find it again. It should still have been there. Well, the two of us searched for ages, all around where I thought it should be, but with no luck. Since the bush was very prickly, I had to get flat on my belly, as low as possible to try to avoid the needles and crawl into the bushes at every place I thought the gun might be. It was awful! We got all scratched and thoroughly disgusted before we decided that it wasn't there. Remember that during this "carry-on" we little boys were wearing short trousers that came only to our knees.

What I finally figured was that someone had seen me hide the gun and, after I had gone, removed it and took it to the proper authorities. Either that, or I had got really screwed up and there is still a machine gun hidden among some bushes for future archaeologists to find a long time down the road. Anyway, it was a very stupid thing to do and I don't know what my mother would have said if I had walked into the house carrying a great big machine gun. One thing's for sure—I would have got a thick ear!

~ * ~

Looking at a map of Campbeltown nowadays, one could see something new called "Kintyre Gardens" just south of Kintyre Park. There is a possibility that when it was being developed (long after I left the area) that the gun may have been found then, for it was in that general area that I hid it.

There was a lot of different-sized aluminum reinforced tubing around the crash site. Some could be cut into bands to make rings for fingers, and filed to different shapes. I remember getting a piece and making a ring for my girlfriend, filing it so that there was a heart on it. I had to start with a piece around 3/8" (10 mm) across (wide), file the band down to ring width, leaving a section wide enough so that it could be filed to the shape of a heart. I wonder if she still has it (whatever her name was, I don't remember). I also wonder where I got the hacksaw that I used to cut the tubing and the file that I used for shaping; for there was no way that we could possess such "exotic" tools - or *any* tool for that matter!

~ * ~

During summer break from school when I was around 12, I made an extended visit to relatives in Glasgow. I don't know who paid for the substantial cost of the bus fare, as the distance by road was 138 miles (220 km) from Campbeltown to Glasgow.

Mother's brother, James Morrans, an aunt and a girl cousin lived on Steele Street, just a stone's throw from Glasgow Green. This was an awful slum area, just a few blocks down the Saltmarket from Argyll Street (near The Trongate and Glasgow Cross). They lived on the second floor of a three-story-high building, up a winding staircase. An awful smell of stale urine filled the air, as if drunks had used the close *and* the stairway for a toilet.

I soon joined in playing with the local boys. One of our favourite pastimes was pretending to be "big game hunters." This consisted of using a "bow" and a few "arrows." The local boys—the owners—stored them when we weren't using them. A nail was tied to each arrow, similar to our "spears" for flounders only in miniature. Then we would shoot through the broken stairway window at any rats that ventured out in the daylight. The window was just a little higher than the flat roofs of the "back-green" wash houses, similar to those in Campbeltown. The hard part of this pastime was getting out to "the back green," climbing onto the roof of the wash-house to retrieve our arrows so that we could continue shooting. We did kill the odd rat, but mostly we missed. Robin Hood didn't have to worry about any of us doing him out of a job! I think we only had two bows. This meant we had more boys than there were bows, so we had to take turns. Whoever it was who fired the arrows had to go and retrieve them!

Our play area was directly across the street from the Tent Hall, a place for religious meetings. The outside areas of the houses were, I think, worse than the inside, if that were possible, for the street was disgustingly dirty. (At least, Lorne *Street* was clean.) Anyway, this is where a bunch of us were crowded around an old man who was sitting on the edge of someone's front doorstep. The shoulders of his black coat were covered in lice. All of us were close enough to see, but stayed far enough away at the same time. There were so many lice that the black material couldn't be seen under the lice in some places. The woman of that house was trying to get him to move away in case he infected her place. People were telling him to go to the hall across the street. I can't remember what became of him, or if we just got bored and wandered off to find something else to occupy ourselves with.

It would be interesting to go back there even if it was just to see what has become of that area in the years between, as quite a few years ago Glasgow was named the "cultural centre of Europe"! To achieve such recognition, all the aforementioned slums would have had to be torn down and nice buildings put up in their place.

~ * ~

Another story comes to mind. I would have certainly been younger than 13. Hughie Campbell (my chum all through school) and I were walking down the esplanade in Campbeltown, one lovely summer afternoon. We were passing Minnie Walker's kiosk, a little hut which she used for selling cigarettes, candy bars, soda pop and chocolate bars. One of us must have had a few pennies in his pocket, for we stood at the counter waiting to buy something. However, neither Hughie nor I had any "sweetie" ration coupons on us.

No one was there. We waited and still nobody showed up. Hughie and I looked at each other, nodded, and each of us took a chocolate bar to eat while sitting on the grass near the kiosk. When we were finished, one of us must have said that we should get another one. So I wandered over to the counter and was in the process of picking up a chocolate bar when my wrist was grabbed. Minnie had been bent down, lighting her Primus stove to make a cup of tea! She held on very tightly and told Hughie to get into the kiosk beside her, which he did. (Silly boy, he should have run away!) Then she called over two teenage boys, asking them to take us to the police station and she would give them some money. So, there we were, getting marched up the road to serve at least what we thought would be five years' hard labour. Both of us were thoroughly terrified by this time. I whispered to Hughie to run when I said so. A couple of minutes later I yelled, "Go!" and pulled Hughie in the direction I indicated. We dropped ourselves over a wall which was shallow on our side and steep on the other. The teenagers chased us for about a minute and then gave up.

Now, Minnie Walker also had a chip shop that sold deep-fried fish and chips (French fries) in the evening. Even years later I still wouldn't go into her shop in case she remembered me and called the police. *I even dreamt about it at night!* It would have been much simpler if I had just gone into her shop and handed her some money, saying, "Here, this is for the chocolate bar Ah stole from ye, and Ah'm sorry." I'm sure she would have thought a lot more of me; but when you're young you don't think that way.

There were two chip shops in town and the other one made very good chips to take out. We ate them out of a paper "poke" (bag) as we walked along the street. Poor old Minnie's chips were often referred to as "greasy potato soup." But as there was always a big queue at the other place, somebody might say, "Well, let's go tae Minnie's, they're not really *that* bad!" So whenever I wanted to buy some there, I had to ask friends to get mine for me.

~ * ~

Normally a fire would be burning in homes every day of the year, unless it was a really warm day, which meant the temperature was more than 66 degrees Fahrenheit (about 18 degrees Celsius). Warm days like that were few and far apart. Normally, our warm sunny days were around 61 F. or 16 C. Often when I was a young man, the newspapers would say that London was sweltering under an 80 degree F. (27 C.) heat wave; and we would think, 'Gee, imagine being in *that* kind of heat!'

Then there was the dampness. Even in the middle of summer, if you took any piece of clothing, held it close to the fire to warm it up before putting it on, and then immediately put it close to a mirror, the mirror would mist up from the moisture in the cloth!

* ~ * ~ *

My stepfather Bill and mother Christina Moorhead. Mother was under 5 feet tall: Bill was about 6 feet tall.

CHAPTER THREE
The Move to the "High Class Hovel"

Life got a bit better when I was thirteen. We moved to a slightly better home a little farther up the street. It was still not good, but it did have an indoor cast iron sink with cold running water, in front of a window, plus (now hear this) a bedroom! There was also a gas fixture in what we might call the living room (which was a little larger than before, too). Common terminology for our new place was a "room and kitchen." It would have been called a "but an' ben" on the east coast of Scotland! (Those who know the old Scottish song, "Wee Deoch-an, Doris" will recognize the term.) *"...There's a wee wifie* waiting, in a wee but-an-ben*, If ye can say it's a braw, bricht, moonlicht nicht* Then you're alricht, ye ken."*[4] *Scottish words: wee wifie (little wife); wee but-an-ben (small room with kitchen); braw, bricht, moonlicht nicht (nice, bright, moonlit night); alricht, ye ken (alright, you know).

Everything was so much brighter. Although it was still a dump, it seemed like luxury to me—I couldn't believe it was *our* place. The sun shone in the window in the late afternoon, something we had never experienced before. We still had the old oil lamp in the bedroom, but that was okay as my brother and I only used it when going to bed. The gas was obtained by putting a shilling (a coin the size of a quarter) into a slot in a

---

4   Excerpt from Chorus of "Wee Deoch-an, Doris," by Sir Harry Lauder (1870-1950).

meter. When there was no shilling to insert, then the gas died and the oil lamp moved from the bedroom to the living room.

There was a double bed in the living room for our mother. I have no idea where it came from, as we only had one bed before and suddenly we had two—complete with mattress and blankets! Cooking was still done on the coal fire but the grate was a little better model. It had a little door at the side which was really an oven. No one at our home had ever come into contact with an oven before but we soon learned that it came in very handy for drying shoes and socks. We knew that people used their ovens for baking, but my mother didn't know the first thing about *that!* The place for keeping the coal was just inside the house door, to the left. That was very good, for it saved us from going out to the dark landing at night. Immediately to the inside of that was a sort-of storage area, long and narrow. However, I don't think we ever had anything to put into it!

I didn't participate in the move. I don't know to this day if I was even aware that a move was to take place. I went to school from the low-class dump after lunch (?) and returned after school to the not-quite-so-low-class dump! I guess that indicates the extent of our possessions—all could be moved in a few hours. I don't know if my mother had any help in making this move; probably not. I can't even remember if we still had the same old furniture; most likely, for anything "new" would certainly have been noticed.

It was great that we didn't have to use the dark 'close' anymore. We got to our "new" home by going up the (unlit) back (past the clotheslines), via a different wider, open type of close (sounds *contradictory*, doesn't it?), then going up an L-shaped open stone stairway with an iron railing (one of the few that were left as it would have been unsafe to remove it) to where two toilets were (one on each side, for the four families), then climbing an inside wooden staircase.

~ * ~

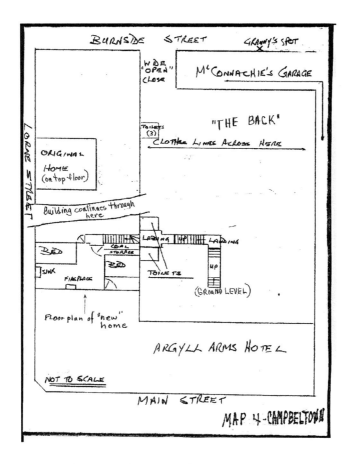

Our "high class hovel," my second home.

Of course, it couldn't be that *all* was improved; that wouldn't do, for we now had to go downstairs and outside a little, to get to the toilet, which was shared with another family. However, having to share a toilet was not too unusual in parts of the town, if not the whole country. Using the toilet was not as good a situation as before. When it was pouring rain, you could easily get soaking wet before you managed to get the toilet door unlocked! If we could have afforded an umbrella we wouldn't have been caught dead with it—something like that was only for sissies! That's another thing that made me laugh. The toilets had a *lock that really locked* with a big key; our doors, where we lived, didn't even have a proper handle! It doesn't make sense—the only thing that would be in the toilet, if you were lucky, would be some old bits of newspaper. I guess it afforded some privacy when doing your business, though.

Our new home wasn't an attic, so we didn't have a sloping wall anymore, plus the two windows (one in the bedroom) were *real* windows that slid straight up and down to open and close, and the sink was just in front of the window in the living room/ kitchen! We still couldn't have a proper bath, but I certainly can't remember any of us complaining. Well, how can you miss something that you've never had? Overall, it was so much better than what we had left behind. Anyway, during the summer we didn't need a bath as we would often go swimming at Machrahanish where there were miles and miles of silvery sands. This was the Atlantic Ocean side of the peninsula and was probably around a "nice" (almost freezing) 50 degrees Fahrenheit (10 Celsius), but it was all right once you dipped yourself in and became accustomed to the water!

If we did lots of swimming it sure helped get rid of the coal dust, too. I can remember going to an aunt's house for a good bath a few times in the winter time when I got a bit older, but mostly I had my wash when my mother was out (this was why she went out). It's likely I was getting too old to have a sponge bath in front of her.

~ * ~

Another change took place around this time, too. I was still 13 years of age when my mother met a man from Larne, Northern Ireland, named Bill Moorhead, who would later become my stepfather. (Perhaps he had something to do with the move.) I can remember Bill staying with us and thinking that we had extra space after our move, and now that was gone. My brother was contributing a little money—just part of his earnings—towards the home as he had started working full time in Kerr's

Butcher Shop. I was putting a little money in also as I was delivering newspapers in the mornings.

One thing I do know is that I wasn't at the wedding, which was held at the Registry Office. I went to school one day and later that day I had a stepfather. (So it seems that funerals, moves and weddings were kept a secret from me.)

I'm still rather undecided whether the marriage was a good thing or a bad thing for my mother; but am inclined to think, "yes" it was better for her. Without him she would have been subject to quite a few years of loneliness when my brother went to live in England and I went to the Royal Air Force for five years, then married and settled down well away from my hometown. With Bill, she now had *respectability*, as she was now "Mrs. Christina Moorhead", and when he died many years later, she was then entitled to a widows' pension from the government.

Life was still not rosy for her even with her "new found respectability" as Bill turned out to be quite lazy. He would work for a little while, then complain that his heart was bad, quit the job that he had, and then lie around doing nothing for months. Then when he did work, he very often got drunk and gave her a lot of verbal abuse. He never did give my brother or me any trouble, though. I guess he knew better. He was pretty smart when he wanted to be. Quite often, he would reminisce about old times in Larne when he and others would make illicit whisky ("poteen" or homebrew). "Poteen boys" he called them. Then he would show me a photograph of a small motor boat that he said he had built *himself* during his younger days in Larne. I was so fed up hearing all that he *did*, that I eventually let it go in one ear and out the other. It seemed to me that all Irishmen bragged a lot about nothing! I don't mean to suggest that we were enemies, though; mainly just the opposite.

~ * ~

Bill taught me quite a few things, among them how to make netting needles that the local fishermen would buy to mend their nets and how to make the best catapult in town, one that was different from any I've ever seen, even today. (A catapult in Britain is called a slingshot in North America.) For all Bill's faults, we got along very well. He certainly wasn't dumb. He taught me things that I never would have learned if he hadn't been around. I don't know what my brother thought of him, for he never ever mentioned Bill to me either positively or negatively. (Actually, my brother never mentioned *anything* to me, period!) I think, perhaps Bill and Archie thought similarly of each other; and I don't think it was complimentary either way.

Regarding the catapult, Bill explained details like making a small hole in the centre of the leather pouch that holds the stone to reduce drag by letting air through while the stone was shooting forward. He taught me to reduce the "Y," leaving just a stub on the bottom leg and cradling the other two legs with the thumb and forefinger. The securing system of joining the rubber to the leather was very elaborate, quite unusual with all the loops of leather that he used. It was actually quite easy to hit a tin can about ten meters (30 feet) away—impossible with the ones my friends and I had made previously. Best of all, Bill emphasized one very good rule: *never* shoot at any living thing!

~ * ~

One day while I was playing with my friends I heard, "Ian, will ye come with me, please? I need yer help." Bill had never said this before, so I didn't know what to expect. It turned out that he had found a small row boat washed up on the shore just outside town. The stern was all smashed away, and I mean *totally gone*. I think it would have been classified as "flotsam." It was perfectly legal for Bill to take it. Wreckage found on the shore was considered "finders keepers."

"Help me to carry it to O'Hara's yard and I'll fix it up; then we can go fishing."

"Sure, Bill," I replied although my thoughts were, 'How the heck can ye fix this when the whole back end is missing?'

But, boy, did he ever fix it! You would never have known that it had been damaged, except for a little different shade where the new wood appeared. The boat was classified as "clinker-built" (overlapping boards). I never did watch him work on it and I've often regretted it. I thought he was just doing his usual bragging. I couldn't believe it was the same little boat when he asked me to help him take it to the water. He had also managed, (how, I don't know) to have two oars and two row locks (oar locks) and "hand-lines" so that we could use the boat right away, plus a rope to tie it up when we weren't using it.

Use it we did! We were never short of fish after that. It was only Bill and I who went fishing; my brother always said "no", although he would certainly eat his share. We would go out in the evening, usually around seven or eight p.m., spend an hour or two at the most, and return home with cod, mackerel, or flounder. There was plenty of fish. In less than an hour, we usually had enough fish to last us a couple of days, plus some for Bill to sell.

~ * ~

Just after the onset of the war, the Admiralty had commandeered an enormous "yacht" (not a sailing boat) that belonged to Lord and Lady Docker. It was renamed the "HMS Shamara" and had its own permanent berth at the old quay. This ship would go out *every night* around eight p.m. and return early the next morning. Nobody knew what she did. Whenever we were going fishing, either Bill or I would check whether she was still tied up, or if she had left the pier. When she was tied up, we fished close to the shore, just west of the small warning beacon, to get us well away from her wash as she passed on her way out. When she was out we fished nearer to the middle of the opening as there were more fish there. (The opening was only about a thousand feet or 330 m. wide.)

One night, however, neither of us had remembered to check. We were right in the middle of the opening to Campbeltown Loch, between the island and the north shore, getting ready to put our fishing lines in the water when Bill asked me if the Shamara was out. I told Bill that I wasn't sure, but that I thought she was out. Big mistake! It was almost dark and we had our lines in the water when suddenly we heard the "swoosh...swoosh" of Shamara's bows breaking the water. In the late twilight, just before darkness fell, we could just make out the massive bow of this big ship heading straight for us. I grabbed the oars and rowed like mad, hoping I was rowing in the right direction. Bill started to shout, "Ahoy, Shamara; ahoy, Shamara" at the top of his lungs. I can still see the dark outline of the ship coming straight for us. She was showing no lights whatsoever (and this was peacetime—after the war!). I continued to row like mad, and all the time Bill was shouting his head off. (Bear in mind that this ship was just slightly smaller than a frigate!)

Bill shouted to me, "Get low in the boat, Ian, as low as ye can and hold on to the sides." He did the same. She missed us by no more than 40 or 50 feet (ca. 16 m.)! We bobbed up and down like a cork, oars in, as we clung for dear life to the edges of the little row boat. (There weren't any rules about life jackets at that time, and we wouldn't have been able to afford them anyway!) Bill knew to keep the centre of gravity as low as possible to help prevent capsizing; that was why he told me to get as low in the boat as possible. After that incident we made double sure that we checked if that ship was at her berth before we left to go fishing!

~ * ~

We've all heard "big fish stories" at one time or another. Well, here's another to add to the list. It was always Bill's policy to ask my mother if she wanted to go fishing with us and she always said, "Nae thanks; ye two go and catch fish and that will please me just fine." At least, she almost always did. One evening when Bill asked her, she just about floored both of us by saying, "Aye, okay, Ah'll go oot wi' ye."

We were at our usual spot between the island and the mainland (yes, the Shamara was already out!), our lines were in the water and Bill had shown Mother what she was supposed to do.

"Go down until the sinker touches bottom, Chrissie; then lift it just off, and then touch it down again ever so lightly so that the hooks are a few inches above the sand. That is where the cod are." (Cod are "bottom-feeders.").

Both Bill and I had landed a few fish and Bill was teasing Mother about coming fishing with us, asking when she was going to catch one. Suddenly she cried, "Ah think Ah've got one; will ye help me?"

"When yer pulling like that, Chrissie," Bill laughed, "and nothing is happening, it means that yer caught on the bottom." (Remember too, that she was a really little lady!)

"Well, if Ah'm caught on the bottom, maybe ye'll fix it for me."

"Okay, I'll loosen it for ye and I'll show ye what t' do so that ye'll know how to do it in future …. Hey Ian, I think she's really got somethin' here, look at the line, it's going all over the place now …. Hey, wow! It must be a dandy …. C'mon over here and give me a hand."

Well, I did go to help, and it took both of us to pull in this great big cod! (Keep in mind, we didn't have rods; we used hand-lines and had to haul the dark, rough twine in with our bare hands.) We didn't have a means of measuring the fish; in fact, no one ever thought of doing that in those days. The only indication we had was that the commercial "fish box" we always put our catch in was about 26 inches (65 cm.) long. (Every fish we had caught before always had fit inside the box.) As "Chrissie's fish" was a full head *and* tail over the ends, it probably measured about 33 inches (84 cm.) or more long. That was some fish!

Of course, this episode totally spoiled our "fish stories" from then on. Any time we were going over the evening's catch, we would hear a little voice in the background saying something like, "Do ye remember the fish that Ah caught, ye two?" That was the one and only time that she ever came fishing with us! (Yes, Chrissie, God bless your heart; you're dead and gone now but I well remember the fish you caught!)

~ * ~

We used to classify Campbeltown as "clannished," meaning that the locals kept mainly to themselves. They weren't aggressive or anything like that; just that it was hard for strangers to break into any of their social circles. Bill was a stranger in town and Irish to boot. Even after a couple of years, he was still an outsider. People would talk to him all right, just that there always seemed to be a barrier that was hard to break down. However, one night while he was "having a pint" in one of the pubs (by himself, as usual), a darts league was taking place and one member of a team hadn't appeared. There were only two other people in the pub apart from the teams, and one of them was Bill. The league members asked the other man if he would complete the team and he declined. I suppose there was no choice left but to ask Bill. Well, he loved to play darts and the rest was history after that. He had the ability to say, "What number do ye want, Ian?" and even if I said "treble 20," that's where the dart would very likely land, almost without exception! He was so good, that after the game that night, every darts team in town was clamouring for him to throw for them. Suddenly he was no longer a stranger. He had "arrived."

Bill had always made his own darts. He would cast the lead onto the sharp point, make the shaft out of bamboo, and was the first (that I knew of) to make and use the folded paper flights that are common today; all other darts I had seen until then had three feathers. It was many years later, well after he died, that I started to see the folded paper type. I'm not claiming that he was the first; but you never know, maybe he was.

Just to give an idea of his accuracy, I will relate one incident. I was keeping him company one night when he was working as a night watchman at Campbeltown Creamery. After he had made his round, he practiced throwing his darts and announced that he was going to try something. (His practice was to secure the flight into the shaft of the dart by binding the four segments of the bamboo together immediately behind it with strong thread.) Anyway, he said that he was going to try to stick the three darts together, one behind the other. It took him about six to eight tries but he *did* manage it. The last dart stuck into the second dart's bamboo and then the weight of the three darts pulled the first dart from the board. If you can't see the difficulty in this, just think of trying to propel a dart into an area less than the size of a shirt button from eight feet (about two and a half meters) away! I'm glad I was there to witness this, for if he had told me about it later, I wouldn't have believed him! (Just like the little motor boat!)

~ * ~

My first "real" job was in Hoynes' Bakery when I was thirteen, just a few months before I left school, working part time. I started at five o'clock in the morning and worked until eight. Along with their morning tea, the Scots, for whatever reason, had to have their morning bannocks, (a thick, flat sort of cake made from oatmeal and barley, and initially baked—hundreds of years ago—on hot, flat stones). It was named after the town of Bannockburn. Robert the Bruce, King of Scots, fought the battle there that won independence from England in 1314! Some Canadians think of bannocks as North American aboriginal origin. Actually, the recipe was taken to Canada by Scottish trappers more than 250 years ago, although the latter bannocks seem to be made mostly of white wheat flour and not oats. The Scottish version is a lot more wholesome.

By finishing at eight, I had lots of time before heading to school for nine a.m. I did this weekdays. Then on Saturday I would work until noon. It was against the law for me to start work so early. My boss advised me of this and said I could get the job as long as I kept quiet about it. I guess he was getting me very cheaply, but I didn't care. I have no idea today just how much he paid me, but at least I was earning more money than I would have without the job!

There were two other bakers in the bake-house apart from the boss. One of them, named Chuck, kept telling me that I was stupid, and then would laugh and joke with me at other times. I took it for a while, until one day I got angry and told him that he was the stupid one, and not to call me stupid anymore. We got along well after that. When I got out of school later in the day, I went to work for an hour at Kerr's Butcher Shop where Archie also worked. This was next door to the bakery. I rode the delivery bicycle around town delivering meat, plucking chickens or cleaning up. This extra money certainly helped us. One really good thing about it was that I didn't need to go shoveling coal anymore, or to "the thinning."

The normal age for leaving school during my school days was fourteen. (Absolutely no correlation can be made when comparing that with today's education schedule. Many years later, in Canada, I was able to sit the GED test and passed. I earned my grade 12 equivalent certificate before "preparation" classes were even offered!) I reached fourteen shortly after the war when things were not quite back to normal—whatever "normal" was. Shortly afterwards, the school-leaving age was raised to 15 (now 16), and certain tests and "O level" certificates were also introduced. No such certificates existed before this. School was all over for me so I really didn't care what was being brought into the schooling. When schooling was over, that was it; you were sent out into the world to find a job of some sort. I was no exception.

In North America, there is so much pomp and ceremony and so much money expended, you would think that Queen Elizabeth II was getting re-crowned in every high school! Along with all the fussing, the youngsters are really excited and looking forward to a future of ruling the world. It's too bad that many of them end up working in a shop or driving a truck at little more than minimum wage. Look at the difference between the education in North America and in Britain, where a student leaves school at 16 with a school-leaving certificate, and if more is desired, he/she can then take another year to get what is called "highers" and if more again is wanted, then the final year of "highers" is taken. I don't know how each compares as to quality (or level) but, if a comparison was to be made between the high-school certificate in North America and the certificate issued to the 16-year-old student in Britain, I'm fairly confident that the British one would win.

~ * ~

I felt it was *big-time* for me at 14 when I said "goodbye" to Campbeltown Grammar School. I got an 8-to-5 job with B.I.C.C. (British Insulated Callanders Cables). There I worked with surveyors who were laying out the countryside for the distribution of electricity, serving as a "chain man," using a light chain 100 feet long during the measuring of any land surveyed.

It was quite ironic, really. Here's me, traveling all over the whole Kintyre peninsula, helping to measure out and profile the land on a graph so that all the farmers would have electricity, and "yours truly" was living in a slum, *inside* the town, that barely had gaslight!

For about two months, I was satisfied in that job until I heard that there was an opening for an apprentice in John McCorkindale's Blacksmith Shop. I found out where the owner lived and went to see him in the evening. It turned out that he was not a healthy man and hadn't worked in the smithy for a few years, leaving the running of the shop to a blacksmith and his illiterate son. It seemed that the boss could see no future in having the blacksmith's son there (who couldn't read or write) so he decided to replace him. I was the replacement! I didn't know the story until after I had started work there, and found out that the father blamed me for his son having been fired. This man, Peter Tainsh, was a huge, uneducated, filthy-minded drunkard. He made it his duty to make sure that my life was as miserable as possible.

I was the striker (apprentice), which meant that it was mainly my job to swing the sledge hammer to draw out the white-hot metal on the anvil. If you ever have the opportunity to watch a blacksmith and his striker at work (the trade has died, so this will be unlikely unless it is in an open-air museum), you should see a synchronization of movement between the blacksmith bouncing his smaller hammer on the anvil and allowing it to bounce to just touch a chosen place on the white-hot metal, indicating where the sledge has to strike. The action is continuous, with the striker hitting the exact spot with the face of the sledgehammer that the blacksmith has just touched with the smaller hammer. At the same time that the sledge strikes, the blacksmith's hammer is bouncing on the anvil, ready to indicate where the sledge has to fall on the next hit.

The hammering is continuous, with the striker's sledge doing all the work. As soon as the sledge hits, the big hammer is slightly pulled toward the striker, its head allowed to fall clear of the anvil. Then it is brought around in a large circle for the power stroke. Both of the striker's hands end up together, one sliding down next to the other at the end of the shaft in order for the full force to be delivered. This repeats continuously, one blow about every two seconds, until the blacksmith decides that the steel is getting too cold. When the steel starts to lose its heat, the blacksmith puts it back onto the forge to reheat. It is easy to tell when this is required, as the sledge then starts to bounce (and ring) instead of making a dull thud, becoming very uncomfortable for the striker as the shaft starts to sting his hands.

Every time I struck the sledge hammer for Peter Tainsh, this is what happened; and I couldn't say a thing. I know he would have loved for me to quit. I was the person responsible for his son losing his job, wasn't I?

Shortly after I started working there, a second blacksmith from the town of Oban, just north of Campbeltown, was hired to operate the forge at the other end of the smithy. Theoretically, I "belonged" to Peter, but when he didn't need me, I worked for Bobby. Those were the only times I learned anything, for old Peter taught me nothing. If I asked him a question, it wasn't answered. I eventually got the message and stopped asking questions. What I did, though, was watch what was happening and remember what I'd observed. Bobby was a bachelor, so he always bought his lunch at a local restaurant. When I could afford it, I would buy my lunch along with him and then get answers to all my questions.

~ * ~

Ah, the irony of life! When I was 17, I was attracted to Peter's daughter. How he could have such a pretty offspring was a mystery to me, for Peter was certainly an ugly man. Anyway, Tina and I started going out together. I didn't think too much of it in regards to her father, until one Monday morning. Shortly after I walked into the smithy, Peter roughly shoved me against the wall, lifting me off the floor by my shirt collar with his two massive hands. Peter's face was inches from mine.

"Ah hear ye're gan oot wi' ma Lassie. Well, nae mair o' it. If ye f--- her I'll break every bone in yer f---ing body."

Tina and I remained friends. I had told her (more or less) what her father had said. We hadn't been very serious to start with, plus she knew what her father thought of me. Needless to say, this crude outburst contributed towards the ending of what could have been a beautiful romance!

First known photo of Ian Morrans, age 14, 1946. I'm on the right side, wearing my "American" suit. Friends Ian Brodie (standing) and George McMillan, (left) pose with me. I don't remember, but presume the photo was taken by someone in the Salvation Army.

Shortly after this, Peter's wife died. I thought it would be good if I went to his home to pay my respects. I had met his wife a few times and found her to be very nice. So I went to the door in the evening, all neatly washed and wearing my suit. (This was one that I'd received in the mail from one of my mother's two sisters who lived in Seattle in Washington State in the USA. I certainly couldn't have afforded a suit at that time! It was really quite nice and fitted me as if it had been tailor-made for me.)

It was Peter who came to answer the door. He looked at me for a few seconds and then I got, "What the f— do ye want?" Then the door slammed shut with me on the outside. I don't mind saying that I thought it would have been much better if Peter had died and his wife were still around!

But one sometimes hears that every person has a soft spot somewhere. Although our shop was termed a blacksmith shop, it was mainly *farriers'* work that we did. (A farrier makes and puts shoes on horses' hooves.) Most of the horses we shod were Clydesdales (big farm work horses), which are indeed massive animals, much heavier and larger than a racehorse. (Incidentally, the Clydesdale is the largest horse in the world, the Shetland pony is the smallest and they both come from Scotland. I wonder how evolutionists can explain *that* one with their "survival of the fittest" theory!)

Before I could put new shoes on a horse, the old ones first had to be taken off. This was done by placing the horse's hoof on my hip, or between my knees, depending on whether it was the front or the rear shoe that was being removed, and, facing backwards, using a tool that looked like a short chisel with a handle welded onto the side of it, along with a hammer, to cut the clinches off the old nails on the outside of the hoof. Then the old shoe could be pried off with a pair of very large pincers.

One Wednesday morning I was preparing to put new shoes on a Clydesdale. I had just removed the right front shoe and was in the act of prying off the left front shoe when suddenly the horse reared, lifting both front hooves high into the air, and taking me with him. This caused me to flip right over; as I had the nose of the large pincers jammed between the hoof and the shoe, and was heaving to loosen it. Suddenly I was lying under the horse, looking up and seeing, in slow motion, two large hooves, away high in the air, and ever so slowly coming down towards me. I somehow managed to roll to the side as the horse came down.

I could easily have been killed, but luckily the only injury I got was a bruised thumb, which the heel of the old horseshoe had just nipped. Peter came over and picked me up, took me over to a seat and told me to sit down. Shortly after, I had to make a dive for the washroom to be sick. When I returned, Peter sat with me on the hearth of the forge, put his arm around me to comfort me, and then told me to take the rest of the day off, go home, relax and take it easy. Hey, this wasn't the Peter Tainsh *I* knew! (Unfortunately, he was back to his old, ugly self the next day!)

Despite his disposition, credit must be given where it is due. Peter was really an excellent blacksmith. The things that man could make with just an anvil, hot steel and a hammer really amazed me. He was ignorant of anything to do with circumferences, areas, diameters and the like. If I had been smart-alecky enough to ask him what "pi" was ("pi" being the ratio of the circumference of a circle to its diameter, which I had learned is 3:1416), he would probably answer that it was something to eat! But bring in a wooden-spoked wheel off a horse buggy or a cart to have a new steel tyre (tire—the rim) fitted, he would measure across the diameter, then along the flat steel bar which he used to make the tyre, make a chalk mark and tell me to "cut it there." I had to cut it *exactly* where the chalk mark was, not a little this way or a little that way! When it was formed round, then *fire welded* in the forge, it was then heated all around to expand it, and finally taken to the platform to be shrunken onto the wheel. It always was a perfect fit—never too tight or too slack! When we did this, we must have had about 20 or more buckets of water on hand all around the special structure we had to use, to quickly cool and shrink the steel tyre after we had driven it onto the rim, otherwise it would have burned the wood. If that had happened, then the new tyre would have been too slack.

~ * ~

One day, I was working with Peter when a lady rushed in to tell us that there was something wrong with her friend, the cleaning lady in the Brethren Hall, the building next door. This friend was lying on the ground and there was a strong smell of gas in the air. Peter followed me as I rushed out to the Brethren Hall. The smell of coal gas was overpowering. Peter and I dragged her out through the doorway to the fresh air. Neither of us had ever heard of resuscitation so there was little we could do. I didn't know too much about anything in those days, but I knew the lady was dead. There was an enquiry held regarding her death and both Peter and I were required to attend.

Peter looked almost comical when he appeared in an old suit that obviously hadn't fit him for many years, for his great big "beer belly" had obviously gotten even bigger. He was an ignorant man, which was obvious whenever he opened his mouth. The person in charge only listened to Peter for a minute and told him to take his seat. When they called me to take the stand, I was kept there a bit longer, answering different questions regarding the unfortunate situation. After going home and getting into my work clothes, I headed for the smithy. Peter was already there and he wasn't short of words right then, complaining (to Bobby) how I had made him look small by "bein' so smart in court and talk, talk, talk."—It was just another day!

~ * ~

During most of this time I was a member of the Salvation Army. My experiences there introduced me to my real avocation in life—making music. Jock McMillan, the father of the young boy whom I previously mentioned having been killed by a vehicle during the war, was back from prisoner-of-war camp. He started to teach the rudiments of music to his eldest son, George, Ian Brodie and me. (All three of us were 13.) I was learning to play the cornet. Pretty soon I was in the Salvation Army band, playing second cornet parts. Sometimes I played tenor horn and occasionally the trombone (self taught)—whatever was needed to help "balance" the band. I kept practicing and then I improved so much that I was promoted to first cornet (playing the melody). I also was allowed to use the hall to sit and play around on the old pump organ, pumping the bellows with my feet. I enjoyed trying to pick out tunes on the keyboard and figured out some chording with the help of a book from the library. I only wished that we had had enough money to allow me to take piano lessons. I also was encouraged to sing solos during meetings. The captain might say, "Ian, why don't ye get up and sing 'O Boundless Salvation'?" (written by William Booth, the founder of the Salvation Army) and I'd do it without blinking an eye. I loved performing.

~ * ~

I remember a sad story from my first time after getting fully involved in the "Sally Ann." Our captains were Mr. and Mrs. Robson. They had one child, a girl about five years old, blond haired and pretty. Not too far from the "quarters" (S.A. equivalent to "manse") was another little girl who was also around the same age. They looked almost like twins, and played with each other, too. One day when Mrs. Robson was in the post office, her little girl was outside and ran onto the road, where she was hit by a car. She was immediately rushed to the Cottage Hospital on Witchburn Road for attention, without the mother's knowledge.

Captain (Mr.) Robson was at the same hospital ministering to a patient. As it was thought that the little girl just brought in was the friend of his daughter, he was sent for. Imagine his horror when he found that instead of his daughter's friend, it was his daughter lying there. Tragically, she was dead.

~ * ~

Some of the Salvation Army band members pose onboard ship which took us to a band weekend in Dunoon in 1948. I'm the one in the middle.

The Salvation Army at that time very much frowned upon dancing and going to the movies. "Worldly entertainment" was considered a terrible sin, as were having a beer or a glass of wine. (Actually, drinking was considered infinitely worse than the movies; you were on the road to Hell!!!) We were taught to keep away from worldly entertainment, and if we were found to be enjoying ourselves in such a manner, we were expected to kneel at the "mercy seat" to ask forgiveness.

I had to kneel at the mercy seat quite a few times, as I would quit going to the "Sally Ann," then go dancing on Saturday nights, start smoking and go to the movies, for a while. All of these things had the makings of a sinner!!!

I was a "natural" at ballroom style dancing, learning it very quickly. All dressed up in my nice "American" suit, I would go to the Bowery every Saturday night and dance with all the nice, young things. The music was provided by Joe Morrans (no relation that I know of) and the four-piece band he had. The music was pretty good.

I also attended a night school to learn Scottish country dancing which is somewhat like North American square dancing. This was also enjoyable and quite different from ballroom dancing.

Perhaps about a year or so would pass while I enjoyed my worldly pleasures, getting quite good at dancing, and then, for some reason (I think it was because I missed playing in the band), I would return to the S. A.—then quit again for a while and go back to the dancing. Then a little later I would return to the band again, so would have to hit the mercy seat once more! Just couldn't make up my mind what I wanted to do with myself! This was when I needed some of that gunnysack for my knees that we had used for the "thinning."

Ian Brodie, George McMillan, and I (with tenor horn) pose onboard ship after a weekend of Salvation Army performances in Dunoon, 1948.

I found the Salvation Army to be quite a time-consuming affair. Here's a sample itinerary: An open air meeting was held on Main Street at 6:30 p.m. every Saturday evening until 7:30, then the band would march back to the hall, playing a rousing march, for a "Salvation Meeting", which lasted until 9:00 p.m. Next morning (Sunday) we all assembled at a predetermined spot somewhere in the town at 10:00 a.m. for an open air meeting. (I can imagine that all the folk with hangovers just *loved* us!) Then at 10:45 a.m. we marched back to the hall, this time playing a hymn suitable for marching, for the "Holiness Meeting" which lasted until noon. We then had to be back for 2:00 p.m. to teach at the Sunday school, then we assembled outside the hall at 3:00 p.m. and marched (playing) to a different part of town for another open air meeting. This lasted from 3:15 until 4:00 p.m. Then we had to be back at the intersection of Main Street and Longrow South at 6:30 p.m. for yet another open air meeting. This lasted for an hour and then we would march back to the hall to another rousing march, for the 8:00 p.m. "Hallelujah Meeting."

This wasn't the end of it. During the summertime, when the evenings were still daylight until quite late, we had late-night open air meetings at the pier-head from 9:15 until 10:15 p.m. both Saturday *and* Sunday nights, marching and playing from the hall to the pier to draw a crowd.

I know, I know! When one considers all the time that was spent in doing all of this, one can easily see that there was method in their madness. There absolutely wasn't any time left *at all* for "worldly entertainment!" I wonder what they do today regarding television, camcorders, IPods and all the movie outlets. Go with the times and accept it, I suppose.

This sort of thinking irks me, for if it was a sin 50 years ago, it should still be a sin today. If it is all right today, then it was all right away back then and we were getting wrong instruction. To my way of thinking, if the S.A. had the same strict rules today that they did 50 plus years ago, very few, if any, except maybe the zealots, would be attending; so I think they *had* to change. Rather like at one time it was wrong for Roman Catholics to eat meat on Fridays; now there is no problem with it! No wonder it is difficult today to get people to be churchgoers.

Here's another example. We had a youth group meeting on Thursday evenings, and each week there was a different topic to discuss. A speaker was chosen every week from amongst the group. Soon it was my turn to pick and talk about a topic. When Captain Robson asked me what I was going to talk about, I said "Worldly Entertainment."

"Excellent, Ian, excellent," he replied. He knew that when I was in the S.A. I didn't go to the movies, dancing or take part in any worldly entertainment, for I'm a firm believer in either being one or the other—no half measures. (Which probably means that when I wasn't active in the S.A. I was a *real* devil!)

Come the Thursday evening during my talk, I said something like, "However, if ye feel *in yer heart* that ye are *no* committing a sin by doing these things, by all means continue. It isn't until ye *know* that a sin is being committed that ye do actually commit one."

Well … you should have heard the uproar I got later for my efforts!

"What d'ye mean by telling them that it was *okay* to sin? Who gives *ye* the right to be judge of what is right and wrong ….?"

I couldn't, for the life of me, get it through to the captain, that if a person goes through life with no religious education, then that person is not held accountable to God for rules *made by man* as to what is right and what is wrong.

~ * ~

It wasn't too long after they had lost their little girl that the Robsons ceased being Salvation Army Officers and he took a job in town. He was well educated, a very good musician and played the organ in Castlehill Presbyterian Church. I'm sure no one can fault him or her for leaving. Here they were, in their mind serving their Lord, and such a thing happens. That's enough to shake anyone's faith. (They both returned to being S.A. officers after about two years' absence and took charge of the corps at Campbeltown again for a further two years. This was after I was accused of pilfering. I went to the Royal Air Force shortly after this. The accounts that I describe below were not during their captaincy.)

~ * ~

Shortly after the Saturday evening meeting started, I would be asked (sometimes), to go to the local pubs (bars) to sell *The War Cry*, an S.A. newspaper. This was done without fail every Saturday night, but usually by different people on a rotating basis. Anyway, this one night, I went on my rounds and, as luck would have it; I seemed to meet with a lot of drunken men. Now, it was my own belief that it was wrong to sell a paper to a drunken person. So what I would do was to fold it up, put it in his pocket in the hope that he would read it the next morning when he was sober and start to lead a Christian life. (I think I missed my calling. I should've been a saint!!! No, forget it—I think you have to be *dead* to be a saint.)

Needless to say, with meeting so many drunks that night, the money box was a little on the light side and all the papers were gone. When I returned to the hall and the S.A. captain opened the box and only a few coppers fell out, she turned to me and asked where the rest of the money was. I told her that I had met a lot of drunks and had put the paper in their pockets. She wouldn't accept this and insinuated that I had pocketed the money. So that was me, quitting again! A few weeks later she was at my door and apologized. I went back, but this time I didn't have to hit the mercy seat!

There was one rather peculiar situation that occurred every Saturday night when it was my turn to go around to the pubs; well, to at least one certain pub. A schedule was always made out and sometimes the way it landed, I had to go into the pub on Shore Street. In this pub there was always one certain large man who would put a hand on either side of my waist and pick me up, lifting me onto a table. (Not a difficult move for him, as I barely reached 5 feet 5.) Then he'd take *The War Cry* papers and the collection box from me, and order me to sing "The Old Rugged Cross"—all three verses. (There are actually four verses, but we only sang three.) Meanwhile, the man would go around to all the people in the pub with the papers and the money box, distributing and collecting. It's no wonder today that I still remember all of the words to that song by heart, fifty plus years later! *"So I'll cherish the old rugged cross 'til my trophies at last I lay down. I will cling to the old rugged cross and exchange it some day for a crown."*[5]

~ * ~

A few months later (I would have been around 16 years of age), it was getting close to what was called "Harvest Festival." That was the time when just about all of the congregation went around different areas of the town to put envelopes in people's doors, and call back in a few days to pick them up, looking for donations. I had an old bicycle by this time. (I can't remember buying it, so it must have been given to me.) I stupidly volunteered to cycle up the countryside to the different farms. (Later you'll see the rule "don't ever volunteer.") This was quite an undertaking, for it meant working in stages.

When I had completed my project, I took my sealed envelopes to the captain so that she could open them, count the contents, and write the amount collected into a ledger. It was obvious that she was expecting much more than she got from me, for she then, again, accused me of taking some of the money. Well, that was it for me as long as she was the officer at Campbeltown! She did come and apologize again, but I had had my share of her mistrust. I didn't go back until Captain and Mrs. Robson returned. (Meanwhile, I was able to "go to the Devil" and also to "the dancing" on Saturday nights!)

I found out later that there was a time, long past, that someone had stolen a collection box from the S.A. hall with

---

5   Excerpt from "The Old Rugged Cross," by Rev. George Bannard (1873-1958).

money in it, and it was insinuated that it had been my mother. (I can't say whether she was innocent or not, but at this late date, and considering that the money was *for the poor*, if she did take it, I would say "Good for you, Chrissie!" for we were "the poor.") If she *were* guilty, then she must have been desperate! Hey, maybe it was to buy a paper airplane for a howling brat!!! But I would like to think that it was someone else who took it.

~ * ~

There was another humorous situation that arose at the Sally Ann. The man who played the big drum in our band was called Willie Lafferty; and he was a little simple-minded. He was all right on the drum, for he could keep the beat quite well. (Now, remember that the man who plays the big drum is away at the rear of the band.) One Sunday morning just after we had finished our open air meeting we were marching back to the hall, along Kirk Street, playing the hymn tune, "Duke Street." We were approaching the main intersection as the local pipe band was marching *down* Main Street and they were playing, too! If we had continued, we would have walked right into them. The bandmaster, Willie Anderson, stopped us from playing and marching until the pipe band had passed. Then we proceeded across Main Street immediately after that. As we were almost over, about to enter Cross Street, the bandmaster told us to start playing again, saying loudly, "Ye can start now Willie." We waited for the drum beat that would get us all started at the same time—"boom, boom, boom, pause; boom, boom, boom, pause." Silence. The bandmaster said it again, this time a little louder. "Okay Willie, start." Still silence. We all turned around to look for Willie and found that he wasn't there!

All of us then ran back to the intersection and looked down Main Street towards the pipe band. There was Willie, marching merrily behind them! He had *joined the pipe band*, beating his drum as happy as could be! Well, we all broke up, laughing so hard we cried! I'm sure there wasn't a dry eye amongst us! And, of course, someone had to go and rescue Willie. Now, that was when we *really* needed a movie camera. I don't know which would have been the funniest—Willie joining the pipe band or all of us scampering back to the intersection to look for our drummer!

~ * ~

During the war years and for quite a few years after that, the people of Britain had to use ration books to obtain most types of food. Everything was scarce: sugar, butter, beef, fruit, petrol, eggs, oil, chocolate, clothing, cigarettes and lots of other things. It was not too bad for anyone who was a regular customer in a certain shop, for then they got preferential treatment. For instance, a smoker would get a better brand of cigarette from his tobacconist for being a regular; but someone else not "in the know" would get "garbage" cigarettes like "Pasha" which smelled like they were made from camels' dung (maybe they were!). The same held true for all the other shops and other goods. We, the underprivileged, had no such good fortune. We didn't have much money, if any, which meant that we weren't "regulars" in anyone's shop, so couldn't be in anyone's favour. This meant that when we were lucky enough to be able to buy, say minced beef (hamburger), for example, then it was the fattiest beef that was ground up for us (the junk that they couldn't, or wouldn't, sell to the "better class" folks). The better meat was kept for the regulars, but we didn't get it any cheaper. Minced beef, by the way, was ground up by the butcher when it was asked for, not in bulk like today, so that made it easy to "pass it off" to us. We never did think of complaining—*we* were having meat!

~ * ~

Campbeltown pals at Army Cadet Camp, Dunoon, 1948. I'm in the middle in the left photo and on the extreme right in the right photo. My buddy, Ian McKenzie, is in the front row, left.

During those years I was also in the Army Cadets. It was something to do midweek, especially during the winter. There I was issued with a tartan kilt that I just loved to wear—it meant that I was truly Scottish. (Well, didn't it? Now I wear my own kilt any time I get the chance.) Good job we didn't wear anything under the kilt, for I didn't have any underpants anyway! I can remember older cadets checking us to be sure we weren't wearing anything underneath.

I held the rank of corporal and as I got older, I enjoyed putting the 14 and 15-year-olds through their marching sequence when they were on the parade ground. I knew all the commands and when and how to give them. This helped me to get some better privileges a few years later during my early time in the Royal Air Force.

~ * ~

Even after we'd moved away from the original slum, "Auntie Mary", Ian and Rita continued to come to live in their slum every year as usual for the two months' school break; although, by that time, the kids had both left school. Rita was getting to be quite pretty and we went out together fairly often. It was on such occasions that I appreciated our old dark close which was dark even during the day. Well, it was nice and dark for "late light" summer nights with Rita—very good for snogging (kissing and cuddling) before she went upstairs. I would light matches so that she could climb the stairs, just like the old days. (By the way, our "old" slum home and the one next to it remained empty from the time we moved out, so I guess no one else wanted them.)

One day when they thought I wasn't around, I overheard my mother and "Auntie Mary" talking. Rita and I "courting" was the subject of the conversation. Specifically, they wished that Rita would find someone else to go out with. I heard Mother say that she hoped nothing ever "happened" between Rita and me; meaning that they hoped Rita didn't get pregnant! I couldn't believe what I was hearing and naturally pricked up my ears to listen more.

It turned out that Rita was the daughter of "Auntie Mary" and *my* uncle John Morrans (my mother's brother), which meant that we were really *first cousins*. I knew enough at that age that first cousins were not encouraged to marry, so I stopped taking Rita out (not that I had *any* thoughts of marriage!) When she asked me why, I was very silly and told her. Of course she *would* have to challenge her mother about it, and then it bounced right back to me. I was given a severe dressing down (by both mothers), which I still remember to this day. (As far as I know, Rita now lives somewhere in Australia.)

~ * ~

In early 1950, I was fast approaching 18, the age at which all young men in Britain were conscripted into the armed forces. Being in the blacksmith trade, I was associated with agriculture, and that meant that I could apply for an exemption. Work was still really tough with Peter Tainsh. The other young blacksmith had quit and moved back to Oban, so that left just him and me. When the time came for me to apply for the exemption, I thought about it for some time, and then decided that I couldn't take any more of Peter.

I applied for the Royal Air Force, even though other young men in Campbeltown had either gone to the Army or the Navy. Eventually I passed my medical exam and was accepted. This delighted me considerably, because it was a little more prestigious than either the Army or the Navy. I didn't know of anyone else in our town that had been chosen for the RAF shortly before me, so it seemed like a compliment. Little did I realize how drastically it would change my life!

* ~ * ~ *

Airman Ian Morrans, Royal Air Force, 1951, RAF photo.

CHAPTER FOUR
Off to the Royal Air Force

I got a letter from the War Office telling me to report to RAF Padgate in England. Enclosed was a travel permit which enabled me to get there at the expense of the government. When I arrived, I had to complete lots of forms, was given more medicals, had more tests to do and was issued all the gear that I was supposed to have.

This was where I first had full use of electric light switches, apart from the Salvation Army Hall! Down is "on," Up is "off" (opposite from North America). I've always thought the British way made more sense, as it is easier to put the switch down on entering a dark room than it is to flick it up! And I found those things called "showers" just great! I also found the food great (although some of the spoiled brats, the ones who used to get fed at home, didn't agree with that description!) I was getting fed four times a day if I wanted it. Most of the other guys described the food as "lousy", "good food wasted" or exclaimed, "The cook should be shot." I, on the other hand, could see nothing wrong with it. It was a lot better than what I had been used to. I thought, 'Hey, this Air Force life is all right!'

Then I was carted off to RAF Weeton. With me away from home and my brother off to work in England a month or so earlier, my mother and Bill were left with the house to themselves. I wondered just how Mother would make out for money, so I applied to have a third of my wages mailed to her every week. It wasn't much, (seven shillings and sixpence or $1:08, leaving me with $2:16 to last all week) but the money situation wasn't a great deal different from when I was younger. My motto was still, "a little is better than nothing."

RAF Weeton was the equivalent of modern American "boot camp." We called it "square bashing." Here we encountered a sadistic drill instructor (DI) by the name of Corporal McLaughlin, who gave all of us an extremely rough going over. (And all the time before that, I thought Peter Tainsh was bad!) The other drill instructors were not good, but McLaughlin was absolutely horrible. (The verbal description given by an officer about drill instructors a little further on is very appropriate.) I often wondered afterwards if anyone had ever got around to murdering him, or at least, to giving him a good beating one dark night. It wouldn't have surprised me one bit.

While I was experiencing this "square bashing," I was in a little bit better situation than most of the lads. My time in the Army Cadets helped me quite a lot, as I knew about "ground combat," could also do things like strip and assemble a rifle and a Bren gun and talk about hand grenades. I had cross rifles on my right sleeve which meant that I had "marksman" status on the rifle range. (I got sixpence extra a day for this—full seven and a half cents. Wow!) I was often asked by the DIs to take some of the more backward recruits aside to teach them how to do drill according to the commands. It made things a little easier for me, plus I didn't mind doing this as it got me away from other tasks.

Two pieces of sound advice were given to me before I ever went near the RAF. First was, "don't volunteer for anything, *ever*" and the second was, "if you can get into a band, then do it."

While I was at the Recruit Training Camp, we were all lined up ready to go on parade one morning when an airman came up to the drill instructor and said something to him. The instructor turned to us. "Are there any musicians here?" he asked. Five of us put our hands up. "Okay, you five go with him." We went along quite happily, got to the Officer's Mess and there we had to move a piano from one place to another. So much for not volunteering for anything! You had to be wise as to how questions were worded!

The Flight Sergeant (whom we called "Chief") in charge of the drill instructors was delighted with me. I guess I was his prodigy. He would take me aside to talk to me about certain things and I didn't realize at first what was taking place. *He was priming me to apply to become a drill instructor!* I was delighted with his confidence. I decided "Yes", I would do it; I would become a drill instructor. The only snag was that to be accepted, I had to become a "regular," meaning I had to sign on to be a full time airman, not just a two-year conscript. That meant a substantial increase in pay. First, because regulars got paid more, and then a drill instructor also held the rank of corporal.

This "signing on" had to be done at RAF Bedford. There were a lot more in-depth tests to be completed than what I expected—tests which were designed to let them know exactly what made me tick, you might say. Then after a few days I had to go for my interview with a Careers Officer. There I was informed that the results of my tests indicated that my career should be technical and mechanical, that I would be wasting my time by being a drill instructor. He explained to me that each "career" had a number and that "drill instructor" was *awaaaay* down at the very bottom, and the lower the number, the less the pay. I can remember him saying, "You really don't want to be an *animal*, do you? Everybody looks down on those people. Even when they go to the Corporal's Mess, (club set aside for non-commissioned officers or NCOs) the other guys know that they only got their stripes so that they can shout at recruits and not get punched on the nose. Your tests indicate that you should really go for something like 'Flight Mechanic Engines.' It is at the top of the list, number one—and the pay is better."

Well, hey, things were looking better all the time! The thought of being able to work on aircraft engines had never entered my mind. 'Wow, great!', I thought. So I signed on for five years in that category, and found out that at the end of my recruit training I would be posted to RAF St. Athens to do my trade training.

I wasn't quite through my recruit training at this time, so when I returned to camp and told my chief what had come about, he did not appreciate it one bit. Somehow, the two weeks that were left were not quite as good as before. I didn't get any more favours, and was put right in with all the other guys, who, of course, had figured out what had happened after I told them what I had chosen. They gave me quite a ribbing for the few days that were left.

~ * ~

Then I got one week's leave and was given my travel warrant to go home. From there I was to go to RAF St. Athens in Wales, about 17 miles west of Cardiff, not far from Bridgend. It was hardly worth going home, as it meant two days' traveling time each way. I just got there and it was almost time to head out again. As things turned out, I really shouldn't have gone at all! Here's why. I call this story, "More Than Slightly Sloshed!"

I arrived at our house in Campbeltown at 9:30 in the evening after being on trains or waiting for connections in train stations for a total time of about 20 hours. I found two of my mother's S.A. lady friends visiting her, the three of them huddled around a roaring coal fire.

Noticing a strange smell, I asked my mother what was causing it. She replied, "Oh it's Bill; he's makin' whisky in the wee room." (To us Scots, a "wee room" is a wee bit wee-er than a small room!)

My Irish stepfather Bill was *not* of the Salvation Army persuasion! Very often in the years since he'd married my mother, he'd told stories of how he and his buddies would go up into the hills at Antrim and make poteen, the Irish equivalent of "moonshine" or home brew. Even then I had no idea that it was close to 100% alcohol! I had heard him remark that he was going to make some "one of these days." I'd heard it so often that I was sick and tired of hearing it, as well as many other of his stories of this, that and the next thing that might or might not have been true. Well, when my mother said that Bill was making poteen, I couldn't believe it. I thought he was doing his usual bragging with no truth to it—and that wouldn't have been unusual!

As I sat talking with the women, basking in their remarks such as, "Oh, Ian, don't ye look handsome in your Royal Air Force uniform," my mother tried to get me to eat something. I told her that I was dead beat and just wanted to go to sleep. At this point, Bill came through, said hello to me and asked me to taste his "brew." What he handed me was a glass tumbler with about four or five ounces of clear liquid in it. Not knowing it was so strong, I had almost downed the lot when Bill shouted, "Stop, stop, that's all I've got."

Because it looked so much like water, I had had no hesitation in drinking it. Remember, I was not accustomed to alcoholic drinks of any kind. I didn't know at the time that I was drinking pure alcohol! There was about an ounce left for Bill after all his hard work. It had taken him two days to distill that small amount.

Shortly after this I started feeling a bit woozy, especially as I had virtually an empty stomach. A little while later I said "goodnight" to everyone and took myself off to bed. Stumbling to my room, I stripped off, hopped into bed completely naked and was asleep almost immediately.

The next morning I got up, surprisingly clearheaded, and wandered into the living room. There was a fine, white dust everywhere and I said to Mother, "What is this white powder all over the place."

"Oh, that was ye last night, ye daft bugger," she replied, in the middle of trying to clean things up. "Ye came through here aboot an hour after ye had gone to bed, absolutely naked, walked in between my two friends and *peed* all over the fire."

"You mean that Ah came through here, sleep-walking, stark naked, in front of Mrs. Campbell and Mrs. MacGregor? Me with no clothes on?" I was horrified!

My mother continued. "Ye should have seen the steam that ye created, plus all the ash that went up into the air with it. There was a stifling white cloud all over the house, not tae mention all over the three of us, and Ah think Ah'll be dusting here for a month. We were afraid tae waken ye in case we did something wrong!"

To add to my embarrassment was the fact that both of Mother's visitors were strict Salvation Army believers. I realized that what they thought about my imbibing—my getting fully pissed—well, that was certainly somewhere above the forgiveness level. By then I must have been the talk of the town! Even a long time later, whenever I was home on leave and I saw either of those women on the street and heading my way, I would cross over to the other side—just a wee bit more than slightly embarrassed! Strangely, no one but Mother ever mentioned the incident to me.

Somehow, it seemed, my Salvation Army days were really over. To quote my mother, "Since ye joined the Royal Air Force, ye've gone tae the Devil!"

~ * ~

But I *did* get into a band. One of the first things a new arrival had to do was to go around the various departments to get signatures on an "arrival form," and one of the places marked on it was "band room." Now this doesn't mean that there *was* a band, it was just a line on the form that had to be filled in if it applied, so I had to ask if there was a band. There was!

The next morning I was sitting in the cornet section of the station brass band at RAF St. Athens. I found out that there were certain perks that went along with being in a band. We were excused from guard duty and fire picket duty, plus I got a nice little card that allowed "the bearer" to go to the head of the queue in the mess hall. Most of us didn't abuse this privilege, and the rest of the airmen, seeing us standing in the queue (the badges on our sleeves identified us) thought the better of us as they knew of our little card. The only time we used it was if we were going to play at some function or other; then we would (usually, but not always) be in our dress uniforms and the other lads then knew that it was necessary for us to jump the queue.

It was standard practice to go to the band room every weekday morning instead of to our regular workplace. The bandmaster would rehearse us for an hour. Then we went to our normal duties.

Each bandsman got issued two brass "lyres" (badges) that were fixed to the upper sleeves of our dress uniform, and two cloth ones that had to be sewn onto our work uniform. This was a means of instant identification for anyone in our department, so that our superior officer would know that we *did* actually belong to the band.

I was what might be termed as a "half-decent player," which means that there were many worse than me, but there were also many I would classify as a lot better.

The rule is—solo cornet sits on the extreme left-hand side, front row as faced by the bandmaster. This was the equivalent position of first violin (concertmaster) of an orchestra. So, if one became a member of a *brass* band and he (or she) was a more accomplished player than the resident solo cornet, then that player moved down one seat, and the better player took over that position. This same rule applied to the horns, trombones, baritones, etc. (It was a sort-of status symbol.) Upon my arrival at St. Athens Royal Air Force Station, I was deemed a little better than the airman that had had that seat; so it then became *my* seat. This situation lasted for only a week or so and there was a "new arrival," a Salvation Army bloke from a town in the Rhonda Valley in Wales, who was an excellent player. Boy, it was a treat just to sit and listen to him play! So I was bumped down one, as there was absolutely no comparison between his playing and mine. I was a distant second! But it didn't really make any difference; we were all there to make music. That was all that mattered.

While I was at RAF St. Athens, our band (about 25 of us) played at quite a few functions, among them a duchess was opening a place in Cardiff, and we played the dignitary music for the occasion. (Although I never was one to encourage the "la-de-dahs!")

I spent some time on trombone at that camp as there was a time when we had no one in that position and the bandmaster asked me it I would help out. That meant that I also had to play trombone in the Salvation Army band in Cardiff (which I soon joined) as it isn't good to switch from one size mouthpiece to a smaller one all the time. The S.A. people were surprised when I appeared on the Sunday morning with it. When you play a brass instrument, you develop what is called an "embouchure" (the "reed" you develop in your lips) so it has to remain at about the same size. For a small mouthpiece like a trumpet or cornet the embouchure is relatively small, somewhat larger for a trombone and even larger for a tuba.

~ * ~

I'm not too well up on poetry, but I found out that it was Alfred Lord Tennyson who said, "'Tis better to have loved and lost than never to have loved at all." Well, I think he's full of "you know what." I'm saying this because *love* probably screwed up my early life more than tongue can tell!

While I was stationed at St. Athens, I attended the Salvation Army at Cathay's Corps in Cardiff and played in their band.

While there I met a young lady I fell head over heels in love with. It really was "puppy love." Of course, I didn't realize this at the time and felt it was the end of my world when she tired of me and threw me aside. Cruel death would have been better, I thought!

The lovesick trombone player at RAF St. Athens, Cardiff, Wales.

I still attended Cathay's S.A. because I admired a few people there who had befriended me and I was often invited to different people's homes on a Sunday for dinner. That's one thing about the Salvation Army, they'll invite anyone (even me!), especially someone in uniform, to their homes. There's no excuse for being lonely if you go there to worship.

One Sunday evening the lead cornet player got into a huff with the bandmaster, took his instrument and walked to the back of the hall and sat there. The bandmaster asked if there was anyone who would take over the position and there were no takers. The band had been ready to play a difficult march called "Under Two Flags" and I think that is what caused the falling out. The place was a bit uncomfortable for a few minutes until I said that I would have a go at it. I laid my trombone down and took the offered cornet from the bandmaster. I "fluffed" more than a couple of notes during the performance, having never before seen the cornet music for that piece, but at the end I got a cheer from everybody (and that just *isn't done* during a meeting—*never*!) It also turned out that the cornet player who went into a huff was my erstwhile girlfriend's old boyfriend. I'll bet *I* went down well with *him!*

It was also during the time I played there that the band was invited to Abertillery in the Rhonda Valley to play a concert. I was asked if I would like to sing something, and said, "Sure, I'll try."

The piece I chose was "The Holy City." I had sung it a couple of times before in my hometown S.A. hall. "*Jerusalem, Jerusalem, lift up your gates and sing. Hosanna in the highest, Hosanna to your king!*"[6] For me, this was quite an undertaking, because Wales is known as the "land of song." For a Scotsman to be asked to sing a solo; well, that could be seen as maybe a wee bit presumptuous. It seemed that I "knocked 'em dead," though. Maybe they were just being nice, giving me an "E" for Effort; but the applause seemed genuine. I'm a tenor, and the top note was a healthy "G" which I was able to achieve with no trouble at all. (I can barely manage top F now.)

~ * ~

---

6   Chorus of "The Holy City," words by Frederick E. Weatherly (died 1929), melody by Stephen Adams (died 1889).

Although my girlfriend had rejected me, her mother told me, "Ian, you're far from home and I'm sure your mother would like to see you taken care of, so even if you two are split up, it doesn't mean that you are not welcome at our home. You just keep coming here as long as you want to and I'm sure someone will do the same for my son when he's far from home."

That lady was the best woman I've ever met in my life, *without exception*. Oh, I'm not comparing her to my own mother, for that is like comparing apples and oranges, plus they lived in two completely different environments. I have no idea of how Mom Lewis would have behaved if she had been in the same poverty-stricken situation as my mother was, or how my mother would have behaved if she'd had the resources that existed in Mom Lewis' life.

She was an angel in human form to me. She delighted me *so much* that I took to calling her "Mom." I asked her if that was all right and she said that she felt honoured. (I can honestly say that I have never known a more delightful person. It is with sincere regret that I didn't call one of my two daughters after her. A beautiful name, Lois; Lois Lewis if you must know the full name, originally of Cathays, and last known address was 38 Caerphilly Road, Cardiff, Wales. Her husband's name was Les, and he was a very nice guy, too. Maybe one of my descendants will one day contact one of her descendants and tell them just how fond I was of her. She set an example of how we should all look towards one another. Maybe one of my descendants will even honour my desire and name one of their daughters, "Lois." I can only hope. And if she turned out to be in any way similar to "Mom Lewis" then she certainly would be "A-one.")

~ * ~

Trade training at RAF St. Athens (Cardiff, Wales).
A Hercules engine. RAF photo.

After I completed my trade training, I was posted to a Coastal Command station named RAF St. Maughan, near Newquay in Cornwall, extreme southern England. (It seemed that every place around there was called Saint something or other!) This was just about as far from my hometown as possible that I could get and still be in Britain. I still had the "puppy love" thing in my head and was still certain that life had passed me by.

I must have been fairly capable at my job, for the sergeant in charge of our "wing" would often give me jobs to do that were actually for a NCO (non-commissioned officer—a junior technician and corporal). One time he sent for me and asked me to go up in "D for Dog", which was a Lancaster bomber, (four Rolls-Royce V-12 Merlin engines) as #4 engine was occasionally reported to be running a bit rough. I was to check it out and report back to him. The pilot took me up and did a few circuits (flying round and round). There was no indication of any roughness, and the pilot agreed. A few days later I was asked to do the same again, and again there was no roughness.

A few days later the same thing happened but that time there was a very slight sound that shouldn't have been there (it was also reflected by the pilot's instruments) and I reported this to "Sarge." The aircraft was taken to get checked out. Some time later, could have been one month or it could have been more, the sergeant sent for me again.

When I entered his office, he told me to sit, then he handed me a chunk of metal larger than the size of a clenched fist. "What do you think that is?" he asked. I replied that it looked like a piece of shrapnel from a bomb. It turned out to be one of the bearing caps of the engine we checked earlier. This piece of scrap had been lying still for most of the time in the engine, then it occasionally was picked up, tossed and battered around to cause the roughness. It was amazing that the engine didn't completely break down! Anything other than a Rolls Royce engine would have!

~ * ~

One day we Aircraft Engine Mechanics were all lined up on the tarmac, heading inside the entrance door of a Lancaster Bomber. We were going through the process of learning how to check the exhaust boost pressure by starting each of the four engines. The RAF was willing to teach each engine mechanic how to start the aircraft engines, put them through their paces—but that was as far as it went. We weren't taught the process of throttling back to enable any one of us to hop into a plane and take it up for a joy ride. Landing a plane was a whole different process. The Lancaster had four engines, but they weren't started 1, 2, 3 & 4, in that order—no, if my memory serves me correctly, you would begin with #4 then start #1, then #3, then #2. I had just gone through the system, revving each engine in order and gave my seat to the person behind me. This idiot got his fingers around the four throttle controls and proceeded to rev up all four engines at the same time! The wheels were chocked so it wouldn't move—what it wanted to do was lift the rear off the ground and put the front of the aircraft into the ground!!! The corporal in charge of us got this idiot out of the seat then cancelled the rest of the guys from learning the "run-up" process.

~ * ~

Sometime later, the sergeant told me to walk about half a mile along the airport and taxi a Spitfire aircraft back to an apron (parking pad) closer to the hangars. Why it was out there on its own, I've no idea. I was quite amazed that he asked me to do this, as it was really a job for a non-commissioned officer. I gladly accepted, went my way and found this aircraft right where he said it would be. I hopped into the cockpit and started it up. (A Spitfire is a single-engine fighter aircraft with a V12, in-line engine. It was the aircraft that saved the U.K. during the Battle of Britain in World War II.) Anyway, I got her started and very carefully got her back to where she was meant to be. Gee, I was thrilled to bits—it was a wonderful experience!!!!

~ * ~

Sarge came up to me one morning, "How would you like a weekend in Scotland, Ian?" He said that an aircraft was heading up to RAF Kinloss for the long weekend and that I could go if I wanted to.

"Just fine," I said, "I'll go to my aunt in Rutherglen." (I still kept in touch with "Auntie Mary", Ian and Rita.) I quickly changed into dress uniform, gathered some things from my locker, had to make a call into "equipment" to draw out a parachute, and headed for the aircraft.

On the aircraft was the Station Warrant Officer (known as "SWO," the highest non-commissioned rank) and a couple of others. He and I got to chatting on the flight up, and when we got off he asked me if I had enough money, as he was willing to lend me some and I could repay him later. I said that I had enough and thanked him very much. This man had the reputation of being a very strict individual, so his offer surprised me. I got the train south from Kinloss to Glasgow and a bus from there to Rutherglen.

Rita and I went dancing in Glasgow on the Saturday night at the Barrowland Ballroom. It had been known in the past to have a very poor reputation and at first I didn't want to go there. Once inside, Rita went her way and I went mine. During the evening I was standing in the queue to buy an ice-cream when a young lady said to me that she had left her money in the cloakroom and, "Would ye be kind enough tae buy me an ice-cream?"

"Sure," I said, got it for her and off she went. About two minutes later, one of Glasgow's "tough men" came to me and said, "Ur yoooo tryin' tae get aff wi' ma wumman?" (Glasgow jargon asking if I was trying to steal his girl.) I didn't even get a chance to explain and suddenly I was wearing my ice-cream all over my face. I should have known not to go to that place. I previously had wanted to go to the F & F Ballroom, much nicer and classier but a little farther away. Rita had assured me that "the Barra'land" had changed and that it was all right now. It had *not* changed and I swore that I would never go back there. When I told her of my adventure later, she thought what had happened was quite funny and had a good laugh. I suppose I was fortunate in a way, for in the past it was quite common for a person to get his face slashed with a razor blade in that hall!

I enjoyed the Sunday and Monday. Then Monday night was when I had to get the train north as the aircraft was returning to camp at 10:00 a.m. Tuesday morning. I guess; no, I don't guess, I know I was a little tardy in getting ready to go into Glasgow for the train. When I got into the railway station I was in time to see the back end of the train heading away from the platform.

"Now what do I do?" I asked myself. I certainly didn't have enough money to make the trip south to camp, and I would have to report somewhere for help. I phoned the control tower at RAF Kinloss from the railway station (took me ages to get the number), to tell them what had happened, ask them to pass the message on and also asked how I could get back to camp. I was told to report to RAF Abottsinch (at the outskirts of Glasgow) in the morning and I would be seen to. I spent the night in the railway station and headed to RAF Abottsinch early the next morning. There I was given a travel voucher back to camp, told to report to the Station Adjutant when I got there, and "shook in my boots" all the way, a day and a half of traveling.

When I eventually arrived, I reported as ordered and the Station Adjutant was quite amused. That was until he asked me if I had handed my "weekend pass" into the guardroom on my way in. I know now that I *should have lied* and said, "yes," but I didn't know any better. I said, "No." He asked me, "Why not?" and I told him that I didn't have one! Well, he just about exploded and demanded the whole story. He phoned for the SWO to come to his office and asked him, "Was this airman on the same flight as you?" A few minutes later it was, "Put him on a charge."

On the way out the SWO said to me, "See; if you had taken the money I offered you, you could've made your own way back here and no one would have known anything about it. All you had to do was phone your sergeant."

I was up on my charge a few days later and was given two days' loss of pay. That wasn't too bad, I had expected "jankers" (which, I think, is what is called "fatigues" or "KP" in North America—that job where you seem to be peeling a mountain of potatoes, plus doing special parades in full gear). The worst of it was that I had gotten my sergeant into trouble for allowing me to go north without a pass, and he made sure that I knew that I was out of favour with him from then on. I had lost my "cushy" position!

* ~ * ~ *

A buddy and I wearing summer uniforms, Egypt, 1952.

CHAPTER FIVE
The Stinking Egypt Stint

The Korean War was going on at this time (1951), and I volunteered for duty there. (I hadn't remembered the advice about not volunteering for anything!) I was so sure that aircraft engine mechanics would be needed in Korea and I still had my "puppy love" problem in my head telling me that life wasn't *any good*. That is the actual reason I decided to apply for a posting overseas.

The following is *my conclusion* that love had screwed up my life really well. Boy, could I tell that Old Lord Tennyson a thing or two!

I wasn't surprised when my name appeared on the bulletin board that I had to report for an overseas posting. Oh, I was going overseas all right; they were sending me to *Egypt, Suez Canal Zone*! (Wasn't that just like the armed forces?! If a recruit had been a chef in civilian life, they made him a clerk. If he had been a clerk in civilian life they made him a cook. That was typical of the way things were done. Most likely if I had requested to be posted to Egypt, I would have been sent to Korea! Go figure!)

I got ten days' embarkation leave plus a ticket home and then was told to report to RAF Lytham, St. Anne's; about eight or ten kilometers south of Blackpool. But before I was ready to go home, I developed a toothache in an upper right molar. This necessitated reporting "sick." I had to be ferried by Landrover to the nearest RAF camp, as we didn't have a dentist of our own. The camp was called St. Eval, I think. When I got in to see the dentist (this is an officer I'm talking about), I could tell that he was drunk. This was about 9:30 in the morning! He did a humdinger job on me that makes me burn to this day!

My mouth was terribly sore the next morning so I had to report "sick" again. The medical doctor (an officer also), gave me two aspirins, and never even looked into my mouth. (I came to the conclusion that these people were there because they weren't fit to practice in civilian life.) That wasn't the end of things. I had to go to a certain room where I got an injection in my right arm for—I think—yellow fever. (I *didn't* wonder then why they asked me which was my dominant hand). Then I was taken by lorry to the railway station, to get onto a train for the first leg of my trip home. It didn't take too long to get to Bristol, where I had to change to get the train north to Crew, then change there for the train up to Glasgow, and by bus from there to Campbeltown, a final distance of 138 miles (220 km.) along a winding road (around by Loch Lomond) and up and down steep hills. That leg of the journey took six hours, which included a couple of stops for tea.

When I changed at Bristol and got on the northbound train to Crew, my mouth was very painful. I loosened my tie a little and undid my collar; then I undid the buttons on my Air Force jacket and slept for part of the way. When I wakened, looked at my watch and thought that it was maybe about time that I got myself respectable, I couldn't move my right arm. It had seized up. Apparently this was caused by the injection. When I did try to move my arm, wow, it was excruciating! I couldn't do it. I couldn't move my arm no matter how much I tried. I was in an awful mess. I had to ask if either the older or the younger lady sharing my compartment would help me. The young lady was asked by the other person (her mother) to fasten my shirt collar button, tighten my tie for me and button up my jacket! (Too bad I was in such a mess, for she was quite pretty, about 17. That may have been a good way to lose my "puppy love" if

I had been able to be a bit more receptive.) She also helped to carry my stuff off the train when we got to Crew.

When I eventually got to Glasgow, I found it very difficult when changing trains, for I had my kitbag, a cornet (in its case) that I had recently bought, a fold-up music stand and my side duffle-bag. I had to put them onto the train one at a time, using my left arm because there was no way I could use my right arm. Then, too, the bus depot was quite a walk from the railway station in Glasgow; so I can't for the life of me think how I managed to get there. I must have taken a taxi, for it's the only way I could have done it.

When I did eventually get home, the first thing I did was go to see a doctor. When I told him of my recent experiences, he examined my mouth and gasped, *"Good Lord, he's pulled some of the skin off the roof of your mouth!"*

I was put on penicillin and had a great big bandage wrapped around my face, covering me from my collar to my nose. That idiot dentist had grabbed the flesh that surrounded the tooth and yanked the whole lot away when he had pulled my tooth! It was a lousy embarkation leave, for my mouth was very sore. I had to report to the doctor every day to have the bandage changed or adjusted. Here I was, supposedly there to enjoy myself, looking a little like a partly-undressed mummy. I had to wear the bandage every day of my leave, and wasn't able to leave it off until I headed to RAF Lytham about ten days later.

~ * ~

I was at RAF Lytham for a week, and tried to make up for my lousy leave by visiting Blackpool every night! Blackpool is a "sea-side pleasure resort" on the north-west coast of England and is always geared up for enjoyment. The other "lucky ones" and I were then shipped from there to Blackbush Airport to get on an RAF aircraft for Egypt. The aircraft had engine trouble and we had to go to a nearby hotel to await instructions. There must have been one crafty devil with us, for the word went around that as we were guests of the hotel at the government's expense, we could run up a bill and it would be paid for us. It was *true*; the manager was only too pleased to give us whatever we wanted. We were getting cigarettes, beer and whatnot on the government's tab!

W-e-e-e-ll, all good things come to an end. The plane still wasn't ready the next day, but the manager was told that our spree was ended. How someone had found out what we were doing beats me!

We left for Egypt the next day. The only thing I saw of interest while in the aircraft was the volcanic Mount Etna in eastern Sicily sending its smoke skyward. We had to touch down at Malta for a couple of hours for refueling. Next stop was stinking Egypt!

~ * ~

In my opinion, Egypt consisted of nothing but a whole lot of dirty sand.

Dirty sand – that is all that I could see—nothing else except for a "strange sort of horrible smell" that followed wherever you went. I soon found out the reason—the toilet waste system wasn't what you'd call the best. Besides that, the so-called toilet consisted of a "desert rose" which was just a funnel stuck into the sand that you peed into! (I never did find out what was below the funnel.)

The Suez Canal, started in the mid-1800s, was built by Egyptian slave labour overseen by a French company. However, it had been operated under the protection of the British as a neutral zone since 1888. (After the so-called Suez Crisis in 1956, control of the canal went to Egypt under a United Nations agreement; but I was there in the years prior to this.) From Port Said in the Mediterranean to Port Suez at the Red Sea, British military camps dotted a strip of land that ran alongside the canal.

I arrived at RAF El Hambra and spent the first three days there. After only *one day*, I was put on guard duty. Ian, who was always in a band, *on guard duty?* I couldn't believe it! Adding to my disbelief, I had no idea what guard duty consisted of. The "old timers" who were there longer than I was (probably four days), told me how dangerous it was. One had to be very wary through the dark of night, as when patrolling our assigned area of the perimeter fence, there was a high risk of being stabbed in the back by an Arab. What an encouraging thought!

This was the time just before the "Suez Crisis," when the Arabs were very desirous of having the British return to Britain, away from the Suez Canal Zone and their country (can't blame them), but they hadn't seriously gotten down to hard and fast methods yet. A saying goes, "The natives are getting restless"; well, that applied very well! There had been reports of an odd Britisher being murdered by the locals, but I, being new, didn't know of this until later.

The first patrol of the guard was 6-8 p.m., second was 8-10 p.m., and my first stint was 10-12 midnight. So, here's me, on my first beat, patrolling my section of the fence, not in the least happy about my situation, and thinking that I had just arrived there and could possibly get stabbed and maybe killed.

It must have been around 11 o'clock, very dark all around, when I heard a slight sound behind me. Spinning around I pointed my rifle in the sound's general direction and yelled the usual, "Halt, who goes there?"

Silence. I waited a little, and then continued my patrol, "just slightly" shaking in my boots. With another sound behind me, I swung round again and issued the same challenge, but this time a little louder. Silence again. I waited a bit more before continuing on my way, shaking in my boots even more. All the while, I swung my rifle back and forth in a wide arc, with a bullet up the breach, ready to pull the trigger.

Whenever the patrolling was done after dark, a searchlight at the top of the water tower shone its bright light at intermittent times to all points of the fence, so that an intruder would not know exactly when he would be lit up. It was on the third sound behind me that I swung around just as the searchlight lit up my area. I found myself pointing my rifle at a desert dog, which we called "piards." Boy, was I relieved! I almost fell in love with that dog! You wouldn't believe the thoughts that had been going through my head!

The dog turned out to be quite tame, and when I called it over to me it came with its tail between its legs. I was sure it was hungry and hoping for some food. We had been issued a sandwich to sustain us through the night. I had mine in my small duffle bag. No doubt the dog could smell it. I got the sandwich out of my bag and gave it to him. Then the searchlight was gone and I continued on my way with the dog following me. I didn't really make a supreme sacrifice by giving away my only food; anything that was created in our mess hall wasn't very difficult to part with! (I was already getting spoiled. That's what comes from being invited to different homes for dinners when I was in Cardiff, when I found out what *really* good food was!)

My date of transfer to Egypt was December 21, 1951, and this happened on the 22nd. I was then transported north the following day to RAF Abu-suer, (pronounced aboo-swer) about 10 miles (16 km.) west of Ismailia, a small Egyptian town, just about on the western limit of the British area.

~ * ~

I still get frustrated when I think that we were probably as close as 60 miles from the pyramids, with absolutely no chance of ever going there to see them. All of the country, except for that narrow strip running north and south on the west side of the Suez Canal, was "out of bounds" for us. There was good reason for this, as one would, very likely, be *going on a one way trip*, but not intentionally!

The following day was Christmas Eve. We were all issued two tickets every day over the festive season to allow us to buy a large bottle of beer with each one in the NAAFI (Navy, Army and Air Force Institute), our canteen. So this meant that each person was *unable* to buy enough beer to get sloshed on, right? No, not exactly. Prior to this I had hardly touched any manner of booze (remember that I was usually in the Salvation Army). Our tickets got us large bottles of Heineken, brewed in Holland (around 750 ml. each). Really good and strong stuff it was, too!

There was an Englishman I had befriended when the two of us arrived at Abu-suer. Eddie Williams and I went to the canteen together and bought our two bottles each. I can remember saying to Eddie that there was a Christmas Eve carol service in the canteen hall, and that we should go there when it was starting. I cannot remember finishing the second bottle!

The next day (Christmas Day) was spent experiencing my first-ever hangover; I really wished I were dead! The lads in the billet were only too glad to tell me that, during the course of the carol service, I was up on the stage, telling the station (base) chaplain that he didn't know how to conduct the singing and that I should do it. I didn't believe them, as usually quite a lot of ribbing of the new lads went on.

Previously, I told of the arrival forms we had to fill in and that we had to go to a whole bunch of different departments and get "signed in" when first arriving at an Air Force camp. (There were also departure forms for when a person was posted out.) The first working day after Christmas, I had to pick up my arrival form to get it filled with the names of all the people in charge of the various departments. As I was walking along the main street (beautiful day, sun shining and quite warm for December—fabulous weather compared to Scotland), an officer was coming towards me. I recognized him to be a chaplain. (In Canada, he would be called a "padre.") I saluted him and he returned the salute as we passed each other. Then he said, "Good morning, Ian." I just about fell through the hole in the ground that I wished had been there! I stopped and he stopped, turning around and smiling.

"Well, I trust you feel a little better this morning? I remember when I was talking to you after the carol service that you were Protestant, and that is what my church is, so just go along to my office. I'll be along directly and I'll sign your card where it says 'religion'." *So the guys in my billet hadn't been kidding me after all!* Oh, oh!

I went in the general direction he indicated, found his office and was shown in by a WAAF (Women's Auxiliary Air Force). She looked at me in a strange manner with a hint of a smile on her face as I entered the chaplain's inner office. (Later I found out that she had been at the carol service, too.)

I only had to wait about five minutes, but that five minutes were about the longest five minutes of my life. I wondered exactly what he was going to say to me about my "carry-on." Then he arrived. He took my arrival form, held onto it and then said, "You told me on (I think it was a Tuesday) night that you enjoyed singing and that you were a tenor. Well, our church badly needs decent singers in the choir and I'm hoping that you will join us."

What else could I say except, "Of course, Sir. What night do you meet for rehearsals?" It wasn't exactly what you would call blackmail but ... then he did sign my arrival form! That wasn't all, though. He went over to a cupboard, took out a bottle of red communion wine, filled two wine glasses, put the bottle back and handed me one of the glasses. "Merry Christmas, Ian."

I said about the same, except that in place of a name I had to say "Sir." Then we talked for a little while but he never *ever* mentioned me having made a fool of myself; and that certainly denotes character! Neither did anyone at the church choir, so I guessed that he had told them not to.

~ * ~

Next I headed for the band room. I found out that Joe, who was only a corporal, was the bandmaster (they were generally sergeants or above). He was away on duty at whatever he did through the day, so I had to see him at the end of his duty. That was after I had gone around to all the rest of the departments.

Bandmasters are always pleased to get new members, and usually even more so when told that the new arrival plays first cornet or euphonium. (The latter looks like a small tuba. The music for that instrument can be quite tricky—and also more interesting—to play and requires a better-than-usual player.)

I wasn't sent to work on aircraft for my regular duty. I had to report to the "ground equipment" hangar. This is where all the equipment was kept for the Royal Air Force Regiment (RAFR), a sort-of mixture between the Royal Air Force and the Army. (I had never heard of them before or even *after* that.) The personnel wore the Air Force uniform but with a different insignia on the sleeves, and they did Army things like going on maneuvers. The ground equipment hanger was also where the units needing repairs were sent. (These units were lots of air-cooled JAP engines that were the power source for small generators).

They were about the same size as lawn mower engines, but *not* made in Japan!

It was okay there, with nobody to bother us, and I was in the station band and the dance band, excused from other duties and all seemed fine and dandy.

~ * ~

The next few days passed by all right and then it was New Year's Eve. Four of us had bought our few beers (not that I was going to have very much, I had learned my lesson for the time being!) and intended to pass a quiet evening by playing cards in the billet which was to be my home for a few days until I was issued a place in the tented area. (This was a process everybody went through to get established into a regular spot and it took me two months to get into a regular billet.) The billet housed thirty of us, fifteen on each side, and for most of them it was their permanent living quarters. The four of us planning to play cards were Eddie, (my "English drinking partner"), someone nicknamed "Wimpey," another chap (whose name I can't remember) and me.

Wimpey was an older bloke who used to be a tail-gunner in the "Wimpey" bombers during the Second World War; and I would say that he was one of the lucky ones. Tail-gunners were usually the first to get hit. He had elected to remain in the Air Force even though he had to lose his (air crew) sergeant stripes after the war.

All of the "longtime" men had been issued a rifle and bullets if they wanted them, but neither Eddie nor I had gotten around to thinking of going for ours. Now, there was a young man named Bennet in our billet. The very first time I saw him he was on top of a table doing an excellent tap dance. It also struck me that he looked a lot like Lou Costello, at that time a famous Hollywood comedy film star.

Later in the day (it was already dark outside), we were sitting around a card table at the bottom end of the billet, playing gin rummy. I was just about tight against the wall and facing into the billet. Wimpey was on my right, Eddie was facing me and the other bloke was on my left. All of the others were enjoying a drop of beer and lots of laughing and joking was going on.

As we were getting on with our game, this chap Bennet, who was a little more than half way up the room from us, took his rifle out of his locker. Wimpey, whom all of us recognized as in charge (although he wasn't, but he was a lot older than we were), looked at him and told him to remember the rule that, "if there's booze in the billet the rifles are locked away."

Bennet totally ignored him, so Wimpey repeated what he had just said. Still Bennet ignored Wimpey and continued playing with the rifle. At this point Wimpey half turned toward Bennet, rose, and said, "Bennet, if you don't put that rifle away, I'm going to go up there and take it from you."

"To hell with you," Bennet shouted at Wimpey, responding by putting the magazine of bullets into the rifle and flicking the bolt to put a bullet into the chamber. He then raised the rifle, took quick aim at Wimpey, fired, flicked the bolt to eject the empty shell and put a new bullet in, ready for firing again. All of this happened within about three or four seconds. Fortunately, Bennett had missed Wimpey; but he walked backwards towards the door, saying, "Nobody move or I won't miss next time."

Nobody moved while Bennet walked backwards to the door, opened it by putting a hand behind him, and was gone into the night.

There were soon people who *did* run out though, for there were about 20 of us in the billet at the time. I guess they had heard that it was much better to be a live coward than a dead hero. Someone got in touch with the Guard House and reported the incident. RAF Police came down and told everyone to get out their rifles and go hunt for Bennet. Eddie and I were told to go to the Armoury, get a rifle and join in the search.

Here's me, on New Year's Eve (*Hogmanay* in Scotland, a time for rejoicing), and I was out looking for an idiot—make that an idiot with a loaded rifle! The bells of the church had already heralded in the New Year and I had eventually got to patrolling inside the camp oil compound. As I heard a sound, I crouched as low as I could behind some oil barrels, and then I saw a silhouette slowly walking towards me. "Stop there or I'll fire," I said.

A very shaky voice stuttered, "A-a-a-are y-y-you l-l-looking f-f-for B-B-Bennet? I'm s-s-sent out to t-t-tell everybody that they've g-g-got him." He was shaking in his boots and I knew *exactly* how he felt! I had been there nine days before him.

Later Bennet was court-martialed and was sentenced to two years' detention (military prison). I was declared a material witness and, as I was walking past where he was sitting after giving my testimony, he said to me, quite loudly, "I'll get you, you bastard!" This stayed with me for years, wondering if he would ever catch up with me. (I had never told *anyone* of this threat before writing this. Not ever! Not even my wife, who would have worried herself to death!)

Before the court martial, the RAF Special Police had us replay the scene. They took a string from roughly where the rifle was to the bullet hole in the wall behind me, stating that the bullet had gone between Wimpey and me and had just missed the right side of my neck by about two inches! A little the other way and it would have wounded Wimpey, a little my way and it may have severed my jugular vein. And I was just over a *week* in Egypt. (What next? Hardly a dull moment, eh? I started to think that my life had been rather peaceful with Peter Tainsh!)

~ * ~

My friend Eddie Williams had to go home suddenly, for his mother was very sick. As he didn't have any money, I heard,

"Could you lend me five pounds, Ian? I promise to send it to you as soon as I get home and get paid by the RAF."

"Sure," I said. I loaned him five pounds (more than three weeks' wages at the time)—and I never saw it, nor heard from him again! Just think what that would be worth nowadays with interest!

~ * ~

Collins, a chap who was with me in the ground equipment department, told me that if I wanted an easy, enjoyable time, I should volunteer with him for the transport guard duty to the Catering Corps to pick up provisions for our camp.

"All right," I said. "Let's do it." So, there were the two of us in the back of one of the open-sided lorries, heading out of camp to pick up provisions.

"Make sure you have your rifle loaded when we head down 'Sten Gun Alley,'" he warned.

"Huh? What do you mean 'Sten Gun Alley'?"

"Oh, that is what they call the bit just outside of camp that is lined with palm trees on each side. That's where the Arabs hide and fire at us as we pass by."

It took me about two minutes to determine that this bloke was nuts! And it didn't take me much longer before I found myself just about as flat as possible on the floor of the lorry as my companion fired merrily at whoever was shooting at us! It was then that I discovered that there actually were people who "enjoyed" war, finding it fun being fired at and firing back at them. I couldn't believe it! Hey, listen, it's not that I'm a coward or anything like that, just that I have this inborn fondness for staying alive.

~ * ~

A few days after the New Year, I was told to move all my gear to a tent. This situation didn't suit me. It seemed as bad—maybe even worse—than my Campbeltown slum. I shared the tent with another three guys. Al was on the right on entering, Ken on the left, "Rip" Kirby was straight in from Al's bed and I was straight in from Ken.

I wasn't long in this tent, maybe just a couple of weeks—and for good reason!

One Sunday morning around 10 o'clock, all of us had elected to miss breakfast and stay in bed. I was sitting up in my bed reading, facing the door of the tent. I think Rip, across from me on my left, was writing a letter and the other two (whose beds were each side of the door of the tent) were whispering something to each other.

At first I didn't pay too much attention as to what was happening. Next thing I saw was Al throw his blankets aside and show Ken his erection.

Ken then said, "Bring it over here, Al, and bring your Brylcreme" (a hairdressing cream to use as a lubricant!). Ken then turned himself face down, doggy fashion, and Al went on top of him and screwed him with Ken saying, "Don't shoot your load inside me, Al," (Do you suppose he thought he might get pregnant?) But I'm sure that Al didn't hear him and did it anyway. After it was over, they changed places.

All the time this is going on, I was getting more and more upset, shouting over to Rip, "Hey Rip … Rip … look—look at those filthy bastards; look at them, look at them!" Rip kept telling me to ignore them. When they were both done with each other, they settled down on their own beds as though nothing had happened. This RAF sure was a colourful life!

The next day I went to the SWO's office and asked for a transfer to another tent. He demanded to know why and I just said that there was some friction and it would be better if I moved. (If I had told the truth, both guys would have been arrested.) I got my transfer.

A few days after I had moved, this Ken bloke was walking down the clearing between the tent rows, heading my way. I had the flap of the tent open as it was "make and mend" afternoon (the time set aside for sewing on any buttons or fixing a seam, etc.). When done, the rest of the day was ours. I had finished all my repairs and was lying down. My bed was at the door in this tent, so I was fairly visible even although I had my mosquito net covering me. Anyway, Ken came towards me, pulling the flap of the tent closed as he entered. With a "Hi, Ian," and sat on the edge of my bed. I didn't say anything; just watched. Then he started slipping his hand under the bottom of my net. That was when I told him that if his fingers came any further, I would break every one of them. He hurriedly withdrew his hand and left the tent. He never tried to bother me again.

~ * ~

Also in this tent on another make and mend afternoon, I was fixing out all the stuff in my kitbag. Now, to do this, I usually took the clothes out a little at a time, so as not to upset them too much; but on this particular day I decided that the contents were in a bit of a mess and really needed sorting out properly. So, to empty the bag in a hurry I upended it, letting the contents land on the floor. Something else came out—a scorpion! If I had decided to empty the bag a little at a time I most certainly would have been stung. I don't know exactly just how poisonous this chap was, probably enough, at best, to make me very ill, at worst, w-e-e-ll … but I "deaded'" him right there and then. I figured I had been most fortunate!

~ * ~

Still at this tent, I was standing just outside one day when a voice said, "Ian? Ian Morrans, is that ye?" I turned around to see who was talking to me, and there was one of the MacAulay twins! 'Oh no,' I thought, 'not *this* bloke!'

I didn't mention previously that every year, the MacAuley family came from Glasgow to Campbeltown to spend a couple of weeks during the summer. Now, I don't know why, but this twin must have felt hell-bent to have a fight with me; not just once, but *every flaming year.* A "real, live, knock-'em-down, drag-'em-out" fight! It didn't matter how much I tried to get out of it; it always happened. I also think that he roamed all over town until he found me—didn't want to go back to Glasgow without his scrap. (One time I ended up with a broken nose which, much later in life, I had to get straightened in hospital; and I think it most likely originated from him!)

"Gee, imagine meeting *ye* here!" he said, approaching me to shake hands. We talked for a while and even had a good laugh about the fights we used to have. I have to admit, though, that when I first saw him, the very first thing that flashed into my mind was the thought he might want to have another "go" at me, and I'm quite sure that he *always* won anyway!

~ * ~

RAF Abu-Suer Band on the march. Egypt, 1951. I'm in there somewhere, playing coronet away in the rear. RAF photo.

I mentioned before that at RAF Abu-suer there was an outfit called RAFR which seemed to be a mixture of Air Force and Army and which occasionally went out on schemes (maneuvers). Well, Joe's second in command for the band was a corporal in that outfit, and when Joe wasn't available, this other bloke took over.

For some reason or another, there had to be a CO's (Commanding Officer's) parade one Saturday morning each month; I think it was likely to help the CO keep feeling that he was important. Anyway, the station band always had to be in attendance and play for the "march on," the inspection of the airmen, and then for the "march past." (COs just loved it when a band was there for them.)

Usually, for the inspection, we would play a slow march like "Scipio", maybe the "Grand March" from "Aida", "May Blossom" or something similar. We would keep playing it while the CO was doing his "thing," as the length of time we had to play really depended on how many of the guys were on parade, and how long the CO took to inspect them. We would maybe go through the piece two or three times. The bandmaster would keep an eye on what was happening and stop us playing at the first possible place in the music as soon as the inspection was over.

One memorable Saturday Joe was to be away somewhere (England, I think, on what was called "compassionate leave") and the other bloke was supposed to take over. Suddenly on the Friday evening his outfit got word that they had to go on a "scheme" immediately. (What they did was play soldiers out somewhere in the desert and pretend they were attacking something.) Anyway, it turned out that this bloke came to me and told me that *I* (as the lead player in the band) had to take the band on parade! This was something I never thought would *ever* happen to me, although I knew exactly what to do for I had played in lots of CO's parades by this time. To add to my misgivings, there had *never ever* been an inspection of *any* band at *any* parade I had *ever* been on. In addition, I had never paid much attention before to exactly how the bandsmen were dressed until that Saturday morning.

My misgivings were justified because, when the inspection for the other troops was over, the CO, instead of going to his "march past" podium, *came over to inspect the band*! I had to bring the band to attention, salute, and then follow immediately behind him; and when I saw the state of dress of some of the players, I nearly died. (We were about 28 in this band.) There were blokes who obviously hadn't shaved for a couple of days, others with a button or two missing or dirty shoes. Worst of all, was a chap with *no laces* in his *very dirty* shoes; so dirty, that instead of being black, they were *grey*! All the time I was walking behind the CO, the SWO (but *not* the same from away back earlier) was beside me. Boy, did he ever throw some dirty looks my way, shaking his head at the same time. It was one of these times when a person is inclined to say, "Why *me*, Lord?"

The walk around got completed and the CO then turned to me and said, "Carry on, Airman." That was all that was ever mentioned. I saluted; he saluted; and then he walked back to the podium, ready to do his walk-past salute. Talk about a sigh of relief—mine was probably heard as far away as the moon! (There was a movie some years back, with Lee Marvin, called "The Dirty Dozen." Well, the first time I saw it, and every time after that, I immediately thought of that time at RAF Abu-suer. Whenever I even *saw the title*, it was enough to remind me. We had a good laugh after the parade was over but at the time I sure didn't feel like laughing!

~ * ~

One day I was told that I was to be assigned a bed in one of the billets. I was given the second bed from the main entrance on the left. It was here that I met Billy Russell, a young man from Bothwell, Scotland. One day he invited me to become a pen-pal with his cousin, Mary Fraser, who lived in Motherwell, near Glasgow. As everybody else had their pen-pals it seemed to be the thing to do, so I said I might as well. I wrote to Mary and she replied; I answered; and we eventually exchanged pictures and continued to write to each other. (It never entered my mind at the time that I'd eventually end up marrying her; but that's getting ahead of the story.)

~ * ~

The photos my pen-pal, Mary Fraser, and I exchanged in 1953 while she lived in Scotland and I served in Egypt. Mine is an RAF photo.

When I was established in the billet, I got myself a pet—a real, live chameleon, one of those small lizards that can change colour to suit its surroundings. This little fellow would pretty well stay by the side table of my bed and catch any flies or "mozzies" that came to bother me. It was great—a bug enters and, whoosh, his long tongue would come out and the pest would be gone!

~ * ~

There was also an Arab who tended to our needs (uniform repairs, ironing, shoe polishing, making the beds, cleaning the billet and each bed-space, going for cigarettes, etc.). He got paid 25 piasters (about 75 cents) from each of us in the billet. You could call him our servant. He looked after two billets and was one of the highest paid Arabs in the village just outside the camp. According to Egyptian earnings, this fellow was making a fortune! He wasn't paid anything by the camp, just what we gave him.

~ * ~

It seemed to me that the two main crops that were grown in the region were watermelons and tomatoes. There were probably others, although I can't recall any. During the two-and-a-half years I was there I didn't ever see one modern (engine-driven) pump for irrigating the farmers' fields from the "Sweet Water Canal." The system that was used was exactly the same as what had been used two thousand years ago. It consisted of a bucket on a rope at the far end of a horizontal pole, which was suspended from a vertical pole by a rope holding it up by the middle. The bucket was lowered and allowed to fill with water. Then the other end of the pole was pushed down, making the bucket rise out of the water. Next the end of the horizontal pole was pushed to the side, taking the other end over the land. The bucket was then emptied and the whole process was repeated until there was enough water spilled onto the field to irrigate it. The method was very labour intensive and had to be constantly manned, whereas a pump would have worked away on its own. How the produce was distributed, I have no idea.

There were quite a few street vendors. I often saw people with barrow-loads of watermelons, selling them on the streets of Ismailia. I clearly remember one funny occasion when one "barrow-boy" was established and another bloke came along and parked his barrow directly across the street from the first fellow. The bloke that was already there started shouting at the late-comer; he shouted back and it gradually got worse until eventually the watermelons from each barrow were being used as projectiles, throwing them at each other while the shouting continued. It didn't take long for the street to be filled with a huge mess of burst watermelons!

~ * ~

Then something happened to make the entire situation infinitely worse. The very first rattling of what was to become the "Suez Crisis" began. The Egyptians who ran the water filtration plant walked off the job, leaving our camp with nobody who had any knowledge of how to make fresh drinking water. All of our servants went, too. This was terrible! It meant that we all had to learn how to do our own cleaning again. So, who would you say the military would consider the best people to put to work in remedying the situation? Why, the people who had some mechanical knowledge, of course (including yours truly)!

I had been quite happy where I was and so didn't appreciate being told to report to the water filtration plant to learn how to make good water. It took a bunch of us three weeks to figure roughly how much chlorine and alum to add in the process of making the so-called "sweet water" from the canal into drinking water. That canal was about 10 to 15 feet wide, ran parallel to the Suez Canal (from north to south), and was the place in which the Arabs washed themselves (among other things). A common saying was that, if one of us fell into the canal, it would take seven different injections to combat whatever was in the water!

At the left is part of the so-called "Sweet Water Canal," Suez Canal Zone, 1953. A reservoir is in the foreground, covered with mosquito netting to keep out bugs.

The only reason the water was called "sweet" was because it was fresh (?) water compared to the water in the Suez Canal (which flows from the Mediterranean Sea to the Red Sea—therefore, salt water). Actually, the Sweet Water Canal was anything but *sweet*, in any sense of the word. I had no idea where it originated or where it emptied—didn't even think about it at the time.

(However, I later did some research and found that the canal starts as a branch of the River Nile, heads northeast from Cairo for a bit, then goes straight east to Ismailia—and past Abu-Suer. Then it splits; one arm heads north and empties into the north leg of the Suez Canal, and the other empties into the Great Bitter Lake. This lake is so salty that it is very easy to swim in—the more salt in water, the harder it is for an object to sink. There is also a Little Bitter Lake at the south end of the Great. The proper name for it is "Isma'iliyah Canal" in Arabic. After thinking about it, there is a possibility that it was excavated to supply the workers with fresh water from the River Nile during the time of the construction of the Suez Canal, which opened in 1869. That makes sense, for they had to have fresh water.)

~ * ~

The first water we made, according to the doctor who tested it, had so much chlorine in it that it would have been dangerous. "Maybe it would even have killed a newborn baby," he said. Well, we had to start somewhere, and it was better to make sure that the water was really sterile than the other way around. We could have ended up with six or seven hundred very sick people.

~ * ~

I had been in Egypt about eight months when *all* of the Arabs in *all* the military establishments in the Canal Zone walked off the job. They just quit; that was it. The predicament that ensued was, in my opinion, the fault of the British.

For many years, water filtration plants, power stations and other services were all operated by the Arabs, right from the top to the general labourers, probably because it was the cheapest way to do it. An Arab could be hired for a fraction of the cost of a European, and most likely the thinking of the British was that they would be there for ever, or close to it.

This way of thinking is obviously a mistake for, if the knowledge isn't passed on or shared and there is only one person capable of doing that important job, then that person becomes practically indispensable. There should have been a mixture of

Arabs and Britishers operating the plants so that it would only have created an inconvenience when all the Arabs quit.

Thinking back, I don't know how the Arabs survived after the walk-out, as the British camps were the only means of employment in the area. Maybe it was all engineered by the Egyptian government in Cairo as a means of trying to get back their land (this wouldn't have surprised me). Plus that, becoming sole owner of the Suez Canal wouldn't hurt them one bit. But what I *did* wonder about later was that there was such a gap between the time when the local Arabs walked off at *our* camp and when the rest of them did so throughout the Canal Zone. If it had happened all at the same time, then that would have created a super-duper headache for the British, instead of giving a few of us the opportunity to learn how to keep the entire zone going. We were pushed into learning about water filtration. There must have been others who were pushed into learning about power stations, although that wasn't quite as important as water. The power stations would have been running at the time, therefore it would have been a simple matter to mark down the dial readings and maintain them.

~ * ~

At the water treatment plant which I supervised, Kabrik Forks, Egypt.

Suddenly there was no one to keep the British camps ticking, no one except us few at Abu-suer who knew how to make drinking water; and all of us know how important drinking water is! I don't know what went on with the power stations and how they coped, but the twelve of us (on shift work—three to a shift) at our camp who had learned how to make (half-decent) water were split up and posted all over the Canal Zone, to be in charge of other water filtration plants after the Arabs walked out of the camps they were in at the time.

I was sent to take charge of a place called "Kabrit Forks." At the time I had been loaned to the Army. There I had to teach other people who were sent there exactly how to treat the water. (The authorities actually wanted me to wear an Army uniform so that I wouldn't be the cause of any friction between the Army and the Air Force. I told them "no," I was in the RAF, not the Army). Luckily, there never was any friction between the Army blokes and me.

However, I had come to realize that I had made a *massive* mistake and by then it was too late to fix it. When I was first sent to the water filtration plant at Abu-suer, I should have gone to see Joe (the bandmaster) and got taken *off* that detail. This is something that would have been really easy for him to do at *that* time—even more so, as I played trumpet (lead/melody) in the dance band, and the officers liked to have their monthly dances. All he would have had to say was that I couldn't be spared from the band and they would have replaced me, for there were lots of other people to pick from.

Instead of doing this right away, I was content to play around at the filtration plant, switching with others from time to time so that I could play in the brass band and the dance band and everything went just fine. But once the other Arabs walked off the job, thereby creating an emergency, it wouldn't have mattered if I had been the best player in the *entire country*; making drinking water was far more important than making music! And that can't be disputed. I was far too late in taking advantage of my situation. And I think it was a *major crossroad* in my life and changed it immensely. Now there was no band for me. I was on shift work; all of my orders came from the Army. Wasn't everything just peachy? Humph!

At this time it was possible to get out of the armed forces if your family had money. The length of time still to be served, your rank, and whether there was an abundance or shortage of your trade governed the amount that was required to "buy you out." A common saying whenever things weren't going too well was, "Dear Mother, sell the pig and buy me out." The immediate response from a bystander would usually be: "Dear Son, I'd rather keep the pig!" I couldn't have engineered such a buy out anyway, although it had crossed my mind.

~ * ~

The following is the funniest story in my book (*for me*) or maybe the most disgusting for some others! I like to call it "Jig-a-Jig in the Desert." All I can say is, "You had to be there!"

Kabrit Forks was about the size of one-and-a-half football fields. It contained three very large filters about thirty feet in diameter, pump houses, additional buildings, an office and two very large tents, all of which were surrounded by an eight-foot barbed-wire fence. The large tents were to house the guard of fifty black soldiers from Mauritius, all belonging to the British Army. (Mauritius was an island republic off Madagascar in the Indian Ocean.) They arrived at 6 p.m. and left at 6 a.m. the following day. These men were generally jovial, even though a sniper from the nearby Arab village sometimes took pot shots at them. Luckily, he was a lousy shot and always missed.

One afternoon while I was on duty, I was informed that there were three Arab women at the compound gate who wanted to talk to the man in charge—in other words, me! This was about half an hour after the Mauritius guard had taken over and all who were not on patrol around the perimeter were already gathering outside the guard tent. They stood in a strange huddle, as if they were waiting for something, while doing a lot of excited talking and milling around.

I went to see what the women wanted and found them dressed from head to toe in the standard black Arab women's robes and head covering. It turned out that they wanted me to allow them to enter into the guard tent to sell their "favours" to the men. They were prostitutes! The oldest one (maybe 25) told me, in very good English (which surprised me), that I could have any one of them for free if I would let them in for one hour!

Hey, there were 50 men in the guard. Three women, one hour, that would be—a quick mental calculation told me that each man would get about three and a half minutes each! I told them to go away.

The woman that spoke to me started yelling to the guard, "You wanna jig-a-jig? Huh? You wanna jig-a-jig? He won't let us in!" You can figure it out—the men did! A sudden roar erupted from the now menacing-looking group.

The Mauritians on guard duty weren't happy about my decision, crowding closer to me. The women wanted in—the men were doing an awful lot of grumbling, getting closer, mumbling more—they wanted the women in, too. I realized the situation was becoming critical.

George, my back-up man, said to me, "I think you'd better let them in, Ian."

"I think so too, George," I replied, feeling worried. "It looks like this could get dangerous!" I decided it would be better to go along with the situation than maybe deal with a riot or whatever might happen to us when it got dark. There were only three of us Brits!

Initially, I had been concerned that this might be the time that the "Top Brass" would come visiting, as they occasionally did. Imagine what would have happened to me had they arrived while the fair maids were inside and all the guards were humping them! But now, I thought, 'To hell with the Brass; look after your own skin, Ian.'

"Okay," I said, "one hour only." I told someone to open the gate. You should have heard the cheer that went up from the guard! One minute earlier they were ready to kill me; now I was the best bloke in the world. I would have preferred it if there had been an officer in charge of them, rather than just one of their own sergeants. He, by the way, was first in one of the three lines for his bit of jig-a-jig!

The women lay down on the beds, baring themselves up to the waist by pulling up the black outfits they wore and spreading their legs, revealing all that nature had given them. Each woman had placed a small container beside her bed in which to hold the money. (I don't know how much they were charging—remember, they had told me I was to get it for free!)

The men queued up in three rows, straight back from the foot of each of the three beds. Some at the front of the line were already standing there with their trousers and undershorts down around their ankles, shirts off for maximum effect, naked except for their boots. Their erections were sticking out, firm and proud! It was an incredible sight watching the three bare, brown bums bobbing up and down in rapid fashion, but not in unison! As soon as one chap finished, no more than three seconds passed until the next chap had taken his place!

A camera shot would have been good if I had been brave enough—a movie camera could have earned me millions in those days! Now, I've heard an old, North American expression; and every time I hear it, it reminds me of that evening. "Wham, bang, thank you, Ma'am."

George and I watched for a few minutes, utterly amazed at the spectacle that was taking place before us; then we went to the office to wait out the time. Another Brit on shift with George and me had remained in the office, totally oblivious to the drama that was taking place outside. When we told him what was going on, we all howled with laughter. We didn't go out exactly on the hour, but waited to give them an extra twenty minutes in the hope that it would be all over when we did go out. Then the three of us strolled over to the guard tent. By this time it was beginning to get dark and the compound lights had come on. The situation was "all clear." The women had already left, which I was very thankful for.

As the three of us were talking later in the office, George asked me, "Is there anything that mystifies you, Ian?"

"Sure there is—lots. What are you thinking?"

He volunteered, "I'm wondering how they got here. They surely didn't walk!"

I hadn't given that a thought, but figured he was right, for the compound was out in the middle of nowhere. "Now that I think about it, they probably had their own version of a limousine service," I answered, laughing. "We didn't see them come and we didn't see them go; but I'm sure glad they've gone."

I was also glad I had handled my first "command crisis" in an exemplary manner. One could say I did it with a stiff British upper lip—as opposed to a stiff something else!

~ * ~

There was one more out-of-the-ordinary occasion at Kabrit Forks—this time unpleasant—and I would rather not have to remember it.

We had an Arab construction company from Port Suez in the plant to do some repairs. It was quitting time and all the workers were preparing to get onto the lorry for the trip home. My colleague George came up to me at about the time the lorry was ready to leave the compound.

"Ian, I saw one of the Arabs winding a length of old cable around his waist under his galabiya." (Pronounced "ga-la-bee-ah," the Arab's long, loose cotton gown.) I halted the lorry and the "Boss Man" came out of the passenger seat. Menacingly, he always carried a lethal-looking bullwhip wrapped around his left shoulder.

"What wrong, Umbasha?" (Arabic for *Sergeant*). I assume they thought that by calling me this, working on my ego, they would be treated with favours.

"One of your men has stolen some electrical cable," I said to him in my best attempt at Arabic.

Looking as if he was going into a rage, I thought he was going to abuse me. However, he turned towards the lorry, letting out a cry and then some more shouting in Arabic, but too fast for me to understand. All of the men who had climbed into the back of the open lorry to leave for home came off and lined up alongside it. The boss then took his bullwhip off his shoulder and, in turn, starting at the beginning of the line, touched each Arab on the tip of his nose with it and asked him (in Arabic) if he had the cable. Each said, "Lu" (pronounced as in "luck," meaning "no"). This went on until he was about three quarters down the line. Then a young Arab answered "Aiwa" (eye-wa, ... yes).

I was just about in tears a few minutes later when this young man had skin torn from his back with that bullwhip. All that for a worthless bit of old cable that was of no earthly good to us! I wished then with all my heart that I had never said a word. I stepped in and stopped the punishment—while he was whipping the lad I had been winding up the old cable—so I went to the lad and gave it to him. I didn't know that it is possible to say "Thank you" and feel the sting of a whip at the same time but this young man managed it!!! They say that life is a learning experience. Well, that was some hell of a lesson! Even now, decades later I get a lump in my throat when I think of that young man. I was told later not to feel sorry for him, that the reason he was whipped was because he was caught. So what; he must have also learned a lesson that *he* would never forget, for he got a whipping that he will never forget—even if it was only for four lashes. I know it's one that *I* will never forget. If he had just asked me for the cable, I would have given it to him, for it was just junk. I wished that George hadn't seen him take it and that I had kept my mouth shut!

~ * ~

I must admit that the Egyptians were excellent "removers of property" that didn't belong to them. They had to be—a lot of them lived by their wits in order to make a living. Essentially I found the Arabs to be really nice people. The only reason that they "took stuff" was because they were oppressed by the Britishers and in many cases the situation was, "help yourself or go hungry." Their part of the country was occupied, just like some European countries were occupied by the Germans during the big war; and I'm sure that the locals didn't get any pats on the back from the other locals for being nice to the Germans. It always pays to look at other peoples' problems from a different angle.

My first example happened at New Delhi Army Camp. That time the Arabs took away a whole tent, blankets, sheets, kitbags, shoes and socks while the four occupants were sound asleep. They first of all took the tent away, then the blankets, then the top sheets, then they tickled the soldiers to make them roll over in their sleep, rolled up the bottom sheet length ways, and then tickled them the other way to enable them to take away the bottom sheet from each person. Very clever! The blokes awoke in the morning with nothing—tent, blankets, sheets, all their gear were all gone; and this was *inside the camp* in the tented area where there were lots of other tents. The Arabs were masters at this, and I had to admire their ability. I think that sometimes it was the *only* way many of them had of making a living, so they had to be good at it.

This next example happened at #10 Base Ordinance Depot. That time a fire-brigade V-8 engine and pump, (weighing about a ton) were hauled out *past the guard house* during the night, up a long driveway and across the main road to the Sweet Water Canal. There it was eventually found because it got stuck and the thieves couldn't get it out. Of course, the British guards swore they hadn't heard a thing.

Another example I only know by word of mouth as it happened before I got to the Canal Zone. I must admit it is *memorable!* At that time the Royal Engineers were laying a new pipeline from a filtration plant to a camp. The British Army workers would lay pipes during the day, about four feet under the surface, fill in the trench and then continue the operation, on and on. Then the Arabs would dig up the pipes during the night, but leave the last one buried so that the British would continue to lay the pipes.

This went on, over 20-odd miles of desert, and when the last pipe was laid and connected to the pick-up, a phone call was made to "turn it on." They waited … no water … another phone call … wait … no water. That was when it was discovered that the Arabs had taken the whole 20 miles of pipeline during the nights! I think everybody (well, most of us—probably not the Royal Engineers) had a good laugh over that one!

~ * ~

Riding my Norton motorcycle in Egypt, 1953.

I had my own personal encounters with dishonesty, not all of it from Arabs. The following incident happened in 1952. There was an Italian civilian working in my camp whose name was Vince. I got quite friendly with him. One time he invited me to his place to meet his wife and have a meal with them. A short time before this I had bought an old Norton 500 cc. motorcycle for five Egyptian pounds (most likely around $15), from a chap who was going back to Britain. During dinner, Vince and I talked about motorcycles, as he had an almost new one. My bike was a really old machine, 1939 (and I'm not sure if that was A.D. or B.C.)! He asked me during the meal if I wanted something a bit more modern. I said it would be nice but that I couldn't afford one.

He said that he could get a really good one from Cairo, much better than the one I had. (There were no restrictions on him, as he was a civilian; so he could go wherever he liked.) He said it would only cost me 50 pounds. *That was dirt cheap* for a "good one," so I said that I could probably afford that if I had a few more weeks, as I did have some money saved and would have the remainder shortly. A few days later Vince came to the filtration plant looking for me and then showed me a beautiful 1948 Matchless 350-cc. He told me I should consider it mine and that I could pay him the balance of the money later. There was no "bill of sale" to worry about; he said all I had to do was register it with the RAF Police so that they would have a record of it belonging to me. I can't remember what happened to the old Norton, and it was years later that I deduced that I had been "set-up" in a way, and that the motorcycle had probably been stolen from some poor fellow in Cairo, with Vince pocketing the cash. There never was another invitation to his place and that was most likely because the first one had served its purpose. (I find it terrible—the way people use people.)

~ * ~

One time I bought a nice watch (dirt cheap) from an Arab in Ismailia and found later that it had no main spring, just a rubber band attached to the mechanism that enabled it to run for about half an hour!

Another time, much later, an Arab came to me and whispered, "Hey, Umbasha. You want to buy beautiful, expensive, diamond ring stolen from Cairo?"

"Let's see" I replied. He took a lovely shiny gold ring from under a fold in his galabiya.

"Look here; come," he said, taking me over to the side, and making a tiny scratch at the corner of the glass of a nearby shop window with it.

"Wow," I thought, "a 'real diamond.' "How much?"

He said I could have it for 10 Egyptian pounds. Oh; my heart surely twisted. Here was an absolutely *beautiful* diamond ring—I could hardly believe the size of the diamond—and I didn't have enough money. Actually, I didn't have anywhere close to that amount and said so.

"How much you got?"

"I've only got 50 piasters" (half an Egyptian pound).

Whispering again, "Give me that and tell nobody."

"Okay," I responded, whispering, too. Wow, I had just bought a ring that would get me a lot of money when I got back to Britain. Great!

Of course, my reaction really amounts to *greed* and *ignorance*. Later that evening when I was on the lorry heading back to camp, I noticed that the ring was leaving a green circle around my finger. When I got to camp, I went up to our mirror to scratch it with the ring and it wouldn't do it. I then found out that glass will scratch glass once ... *just once*. The Arab knew it but I didn't. He had to take the 50 piasters as he couldn't "demonstrate" the quality any more – and I'll bet he still made money! I repeat, "Life is a learning experience"! The way I see it today is that I *deserved* to have been cheated. I was the willing participant in the resale of "stolen" property with no thought of the loss to another person. What goes around comes around.

~ * ~

Back at Kabrit Forks, I befriended an Arab farmer whom I often visited, not very far from the plant. I got to know him by buying watermelons and eggs from him. Sometimes I would sit and teach him a couple words of English and in turn learned some Arabic from him. (Imagine an Arab talking with a Scottish accent! It's quite funny, actually. On the other hand, my Scottish-accented Arabic was probably also pretty funny!)

One day this Arab farmer and I were sitting chatting. He was in the process of lighting his "hubble-bubble" (actually called a "hookah"—a water pipe for smoking tobacco with a little bit of hashish on the top!) His 12-year-old daughter, Maneera, was working in the field quite close to us, bent over at the waist. The farmer asked me if I would like to buy his daughter and take her back to Scotland with me. I could have her for twenty pounds he told me, as he lifted her loose, cotton dress (all she was wearing) to show me *more* than her nice bare backside! I couldn't believe it, she just kept on working. I thanked him for his offer, told him that it was a very good one as he had a lovely daughter (she was!) but explained that the authorities would never allow it as she was too young to get married, according to British rules. (Whew, I was sure glad I thought that one up quickly!)

~ * ~

Nothing more of consequence happened at Kabrit Forks. Later I got transferred to another place out in the wilds called "Geneifa," really far out in the middle of the desert. Actually I was transferred to a Royal Engineers camp, next door to #10 Base Ordinance Depot (from where the pump had been stolen). This filtration plant was directly across the Sweet Water Canal from an Arab village.

Close to the middle of the compound was a long disused building except for the end unit, which we used as an office for doing all the paperwork related to the chemicals used at the plant. As this place was out in the wilds, we had to return to camp by lorry at the end of each shift after our replacements took over. (An Army lorry arrived at mealtimes with food for us.) This was a bit of a chore, so I had a good look at the unused building.

The units didn't need much fixing up, so I talked to the lads and it was agreed that we could (with permission) make this place our living quarters and do our own cooking as there was a small cookhouse there. The building was made of brick and that was a lot better than the tents we were in. I put in a requisition for some paint and brushes. It was approved and we got to work. I had a nice room for myself (well, I was the head honcho, wasn't I?) and the other chaps, twelve in all, were two, three or four to a room depending on its size.

This was fine, but only for two weeks. Then an inspection was made by a passing Army brigadier who decided that he would like to see how the water his men were drinking was made. I was on duty when he arrived at our main gate but didn't know of it until one of the blokes was asked who was in charge. The brigadier was taken into my office to meet me. There I was sitting in just a pair of khaki shorts, no shoes or shirt. I excused myself and said that I would put on my uniform.

He replied that it wasn't necessary, as he just would like to see what went on. Then I took him (and his "entourage" also) on a guided tour around the installation, showing him where the water came in, how it was pumped to the filter and then purified (underground chemical room for the chlorine and alum), then to the pumps (two high-speed diesels in one underground engine room, and four electric in another underground room; two of them for canal water, two for good water) that pumped it to the camps. I felt quite awkward taking him around while being improperly dressed.

The first pumps I showed him were the high speed diesel engines. He studied them pretty closely for a while, and then said to me in his "stiff upper lip" Oxford English accent, "I don't see any spark plugs on these engines."

"That, Sir, is because they are diesel engines and are not petrol-assisted for starting." That seemed to satisfy him. (Some diesels are designed so that they can use gasoline for starting—therefore have spark plugs; others are sometimes started with compressed air, after setting a piston to the correct starting point.)

Then I took him to the underground electric pumps with which we alternated. Here he was stumped again. He actually turned to me, after examining the electric motors, and asked me where the spark plugs were.

I couldn't believe it! How do you tell a brigadier that he is a blithering idiot? "Oh, they don't need spark plugs, Sir, these models are self starting." I wondered what his followers thought of him!

Next I had to show him our living quarters. He actually spent more time at this than at the workings of the plant. And I didn't know why until a few days later.

After the visit, WO Lodge, (remember the name) who I got on with really well, sent for me. "Ian, I'm sorry to tell you this after the work your men have done to the building at the compound, but it has to be vacated. I just got word from Headquarters that it has been deemed unfit for human habitation, so you'll have to tell everyone that they have to return to camp tomorrow. I'll arrange a lorry for all of you to bring all your stuff in."

Mr. Lodge himself knew what the move was about, but he couldn't say too much. It turned out that the lame-brained "spark plug" officer had complained that the people who worked at the filtration plant had better living accommodation than *he* had. (I must admit that our living quarters did look very good.) It just shows you how some people think, doesn't it? Not, "Well, good for you lads, the facilities are here and you might as well use them." No, just like spoiled brats (which most officers were, in my opinion), they would instead think, "It's my ball and I'm going home!"

We vacated the buildings, returning to the tents in the camp. I often wondered why it was that the British Army did so well in time of war with such incompetents leading the men. Just think—with really competent officers, we probably could have conquered the whole world!!!!!

~ * ~

Also at this same filtration camp a guard came every night and patrolled the camp until morning. The guard always consisted of Mauritian troops, led (this time) by a white officer and sergeant. We were doing 12-hour shifts, 6 p.m. to 6 a.m. I had just gone on duty when this first class idiot officer (yes, another one!!!) came to me with his "standing orders" clip board.

"I need three men on rotational shifts to guard the filter."

"I'm sorry, Sir; but the only ones here are the three of us to run the plant."

"Standing orders state that three of the camp personnel will be supplied to guard the filter, and that is what I demand."

"Sir, if we have to go on guard duty, I will have to shut down the filtration plant, for we can't be responsible for the installation if we are not able to keep an eye on things."

"You do what you have to, all I know is that this document tells me that three men will be supplied, and if there are three of you, then that is all I will need."

"Sir, this is the first time this has been requested. All of the other guard officers didn't need the filter guarded as it is in the centre of the compound."

"You will do as you are told and supply three men."

"Yes, Sir."

I decided that as I was in charge of the plant, the pumps supplying the Army camps, and also the pumps drawing water from the Sweet Water Canal would be stopped and the chemicals shut off. The whole plant therefore was shut down, with no water coming in and no water going out. There was too much responsibility when no one was able to tend to them. I also decided that I would do the first guard duty from 6 to 8 o'clock (p.m.—which was daylight). I did my first stretch, had my four hours off and had just started my second shift when the sergeant of the guard came to me at about a quarter after midnight. (I'd had a good chat with him during my first stint and thought that he was all right.)

He told me that I was wanted on the phone in the guard tent. I (being a sort-of stickler and maybe just a wee bit of a rebel), told him that I really couldn't leave my post without a stand-in, as his officer buddy may go nuts. He then agreed that he would have to take my place. I went to the guard tent and took the phone from the idiot officer. The caller turned out to be a brigadier from one of the Army camps that we supplied with fresh water, telling me that he had been disturbed from his sleep and informed that there was a severe water shortage at the mess hall. He demanded to know why the water level in their water tower was at an all-time low and dropping rapidly. I told him my story and the air turned blue.

"Put that @$#%&* idiot on at once."

I turned to the guard officer and said, "I think he wants to talk to you, Sir."

Oh, I can *still* relish the moment; it was a treat to watch that idiot officer's face change the way it did, as he stuttered out, "Yes, Sir; yes, Sir; at once, Sir!"

The guard was canceled immediately, but—unfortunately—not without a backlash later. This involved my little white pup, about four months' old, who was actually a desert dog. I had found him just outside the camp gate. He was a nice wee thing whom I called, "Laddie."

Cuddling my desert dog, "Laddie".

Probably about a week after the guard incident, the same officer was on duty (others had been and gone). He came to tell me that I had to get rid of the dog, as it was a danger to the personnel. He insisted that, as it had had no injections, it could carry all sorts of diseases. I tried to assure him that there was nothing to fear as it was a very friendly little dog and wouldn't bite anyone. He wouldn't hear this. I was ordered to take the dog away from the general area and he would have one of his men shoot it.

I shot Laddie myself. I wouldn't have anyone else do it in case they only wounded him, requiring another shot. I knew if I did it myself it would be painless for him. Silently I wished that officer "many happy returns" of bad fortune!

~ * ~

While I was at this same place my buddy George (who had been transferred with me) and I roamed around the Arab village of Fayid one *dark* evening (some place we certainly should *not* have been). We had just bought a bottle of cheap brandy and were looking for a corkscrew to pull the cork out. It is amazing how many mud-hut doors we pounded on, trying to make them understand that we needed something to remove the cork! I had learned quite a few words of Arabic, but none remotely resembling "corkscrew."

I guess we didn't realize that, since those people were Muslims, booze was taboo. It was amazing, too, that we were still alive afterwards to tell about it! Anyway, we didn't get a corkscrew and had to end up pushing the cork into the bottle. After we both had a swig of it, George said that he didn't like it. That left me "obliged" to drink it on my own. I didn't like it either; but after spending all that money I wasn't going to throw it away. I can remember getting on the bus to go back to camp, singing merrily so far down the road ... but that was it.

The next thing I knew it was three days later! I awoke, still fully dressed, *on* my bed and was told that *WO Lodge* had told the other blokes to cover for me. I haven't allowed brandy into my mouth since then. Wonder why?

~ * ~

Being stationed in Egypt on our isolated desert assignments, we all found it fairly difficult to keep our morale up. It was a constant battle of wits, trying to think of something interesting to pass the time. We were issued little metal cylinders that looked like miniature oxygen bottles (spray bombs) which were filled with a chemical for use against mosquitoes. If I remember right it was DDT (now banned!). The small end was sealed with a copper strip, which, when removed, sprayed the tent with a high pressure spray, to kill any "undesirables" that were in there before we entered.

We had quite a few of these bottles (now empty) lying around. I had one in my hand when I said to George, "Maybe we can make a model rocket with one of these." (It was a cylinder about three inches long (76 mm.), and reduced like a bottleneck at the "spraying" end.)

George asked me how we could do it. I said that first thing would be to solder three (tin) fins onto the back end so that the cylinder would be able to stand on them, extending beyond it by a few inches. Next we'd take some .303 bullets apart and remove the cordite strips. Then we'd chop up the cordite into little pieces (it was a dull greenish, straw colour, about as thick as a big sewing needle, and a little less than two inches long). I put some into the cylinder, then rammed in some aluminum foil (from the paper in cigarette packages), then more cordite, more foil, more cordite, etc., until the whole cylinder was jammed full. Then I tied some lengths of cordite together, overlapping it a little and tying it there with thread, to make a long fuse, pushed it into the "rocket," got it pointing to the sky, and lit the end of the *ve-e-ery lo-o-ong* "fuse."

George and I ran for cover, and were peeking around a corner waiting to see what would happen when the fuse ignited what was inside. It simply blew up. Boom, gone. A piece of the "rocket" went through the fender of a nearby Army lorry (but we didn't know a thing about that, of course!)

All that work for nothing! Oh well, back to the drawing-board. We decided that I must have used too much cordite and not enough foil. So I made another one, only this time I chopped the stuff up much finer, used a little less, then more foil, until I had about twice or three times the layers as I had before, to retard the explosion. This time I decided on only two fins, bent a little at the ends to make it rotate and help it to go straight, and that I would point it towards the south desert at about 45 degrees. I figured out a launching ramp of dirty sand (which compacted well), laid the "rocket" on it, and with the same sort of fuse, set about launching it. We watched rather apprehensively, hidden of course, expecting to see it explode again. It disappeared in a sudden swoosh. Both George and I walked a long way, about 50 feet apart, in the direction it went, but never did find it. Of course, we were assuming it did go straight; but who knows? We had a good laugh, saying that maybe it went as far as Port Suez and landed at some Arab's feet! Maybe it only went *behind* us by merely a few yards—we didn't think of looking *there*!

~ * ~

About that time, I got a letter from my mother telling me that Peter Tainsh (the blacksmith in Campbeltown) had died. I felt absolutely nothing. Really! She could have just as well told me that a boy on the other side of town had lost his pet frog.

~ * ~

Trying to get a suntan at RAF Abu-Suer, Suez Canal Zone.

Our camp had a very high water tower, about a 120-foot climb. "Yours Truly" decided to scale it one nice sunny day with a blanket over my shoulder. I laid the blanket on the roof of the tank, stripped absolutely bare and laid down on my back to get "nicely tanned all over," or so I thought. I was young and dumb enough at that time to ignore the fact that I'm a typical "Celt" with a very ruddy complexion (described as "fresh" on my military papers) and so, a tan for me was next to impossible. Anyway, I laid myself down and almost immediately fell asleep. I awakened approximately two hours later, burnt to a crisp! Ah, that *bloody hot* Egyptian sun!!!

Unfortunately, I couldn't report sick to get any treatment because *all* military personnel were classified as "government property." As I had damaged myself in my effort to get a sun tan, I could have been put on a charge and court-martialled for damaging government property! For days I walked about with my hands holding my pants legs out from my tortured thighs— even my "willie" was sun burnt! It was terrible—even going for a pee was very painful.

(I'll never forget that episode in my life; doubly so because it left me with a condition called "Solar Keritosis" in my later years—a pre-cancerous skin condition for which I now have to have regular sessions where my doctor freezes off the lesions, mostly on my head, with liquid nitrogen. Lately, I've had three surgeries to remove basal cell carcinomas from my scalp, on the tip of my nose and near my eye. Ever since, I always cover myself with long sleeves and a hat when in the sun. Just a wee bit too late!)

~ * ~

Just next to the filtration plant, about 200-300 yards away, was a terrible-looking Arab village (entirely made up of mud huts). Occasionally there would be some sniper fire coming from the village towards the plant at night. No one was ever hit, but it was not nice for the people who were guarding the fence. The village authorities were warned about this quite a few times. Then one day six to eight bulldozers lined up just outside the village.

The Arabs were told to remove what belongings they had and that they had one hour to do it. It didn't take long—not much more than an hour—and there was nothing left, certainly not enough that a sniper could hide behind. After that problem was removed and out of the way, the guard could rest a little easier. I didn't feel the least bit sorry for the Arabs. They allowed the snipers to use their village for cover; did they expect the British to do nothing about it?

~ * ~

An Egyptian village being demolished by British bulldozers to avert snipers. 1952.

While still at the same filtering plant, we slept at the Royal Engineers Depot. Next door was #10 Base Ordinance Depot which had a canteen, a mess hall and an outdoor swimming pool which we were allowed to use. Another couple of RAF chaps and I were allowed to use the Corporals' Mess. Corporals were higher in rank than we were as airmen but, I think, here on an Army base, it was thought there was less likelihood of any friction between the Army and the Air Force if we Air Force chaps used the Corporal's Mess. (It's the same the world over—People can be stupid!)

Anyway, here's us, wanting to go for a swim one afternoon, and while walking along the main road towards the pool, we had to pass the parade ground. There was a platoon of soldiers being put through their paces by the regimental sergeant major, and the three of us stopped to watch.

B-i-i-i-g mistake. I learned later that *nobody* stands and watches people on the parade ground!

"You three, you three on the road.....AY....TEN....SHUN!" What a holler!

Suddenly I realized that he was shouting at us. Didn't he know that we were in the Royal Air Force? The order was repeated and then we realized that we had better do as we were told. I think the three of us jumped to attention at the same time. Next thing was, "Qu-i-i-i-i-ck march." Then we were taken onto the "square" (as it's called), and drilled really hard for over a half hour. That cured us from watching other poor unfortunates getting put through their paces!

~ * ~

One time a lorry was going into the desert to take some of us blokes out for a trip; I think to break the monotony. The driver asked us where we wanted to go. The most common response was "Britain," but he ignored that request. Someone said to go down to the Red Sea, and he said, "Okay, but take your rifles." It was quite a rough trip and while we were down there someone said, "It must have been about here that Moses parted the Red Sea." (I checked later and we had certainly been in the same general area of the Sinai Peninsula, but not on the same side of the Red Sea as Mount Sinai, which is further east.) One of the lads picked up a long stick and waved it at the waters, Moses style! *What do you know?* ... It didn't work. I guess he had the wrong stick!

~ * ~

Another time I had to remain at the filtration plant for a couple of extra hours. I can't remember why, but when I got to the camp, the dinner period was over. I went to WO Lodge who gave me a note to give to the duty cook. (I was *taken* to the mess hall by Land Rover. Hey, I could have stood a lot of that treatment!) When I handed the note to the cook, he asked me if I was in a hurry and I replied that I wasn't. He told me to sit down and he would fix me a meal.

Boy, did he ever! I ended up with an absolutely marvelous tuck-in, by far the best that I ever had without having to pay for it. He said that he was a chef in civilian life and that he rarely got a chance to deliver a really good meal. I surely was glad that I had been delayed. (And here's me saying a while back that the military always put men in the wrong occupations! OK, so I made a mistake in one instance.)

~ * ~

One major difference for us RAF types in those Army camps was that we had to go to an RAF camp to get paid. The nearest camp to us was RAF Shallufah. To get there the lads usually were taken by lorry. I was a little more independent, for I had my motorcycle. Anyway, there was this other Scotsman in our lot by the name of Bob, from Falkirk. This bloke was continually boasting of his dirt bike riding "back home." It so happened that my usual buddy couldn't go with me for one trip and Bob asked me if he could get a ride with me. I always felt it was better to have a companion aboard (it meant an extra rifle) and I said that he could come with me.

The road to RAF Shallufah went south for a few miles and then west on a side road to link up with another road. Then we headed south again, taking us to the camp. So, I thought, 'as this chap likes dirt bike riding, therefore he must like thrills on the motorcycle.'

When I headed on the westbound road, I skidded right across the highway at the intersection, going down a very wide ditch. In the process, Bob came flying over my shoulder, and then I continued up the other side.

As Bob picked himself up off the ground, he told me, and he was really angry, "You must be the worst driver in the whole world!" (Funny thing though, we didn't hear him brag about his dirt bike riding anymore!)

~ * ~

There was an Army corporal at this camp, an Englishman, who always wanted to borrow my motorcycle. I always told him,

"No." One evening he came to me, quite drunk, and asked again to borrow my bike. Again I told him, "No."

Well, the next morning when I went outside to ride my bike to the plant, it wasn't there. I started asking around if anyone had seen my bike and then the "borrower" told me that I had said he could borrow it and, unfortunately, it had caught fire. He said I would find it in the middle of the power station yard. I went there with him and, sure as fate, there it was—a burnt-out pile of junk.

I told him that I was going to the guardhouse to lay a charge and he responded, "Go ahead, it's your word against mine. It wasn't my fault that it went on fire after you had lent it to me." Then he continued with a smirk, "Just see what you can do about it!"

I was too disgusted after that to even bother, deciding it would have been a waste of time to try to get anything out of him. I wouldn't have minded too much if he had offered to pay for the damage, or at least made some sort of offer, but I guess that was too much to ask! He must have wheeled it away for some distance before starting the engine in order to go joy riding or I would have heard it. I later figured that he must have let it fall on its side with the engine still running, letting gasoline leak all over. He was certainly drunk enough for that to happen. I didn't buy another bike, (besides, the Italian, Vince, wasn't around!)

~ * ~

I was passing by a sort of lean-to shed one day and there, lying alongside the inside wall, was what had once been a beautiful wooden canoe. It looked like it had either been run over by one wheel of a lorry or a heavy pole had fallen on it, for the whole centre section had been destroyed right across it. Looking at it, I imagined the job that my stepfather, Bill Moorhead, would have done on it and thought, 'Maybe I could do that.' I found out who it belonged to and the bloke told me that he wasn't interested in it anymore and that I could have it. He also gave me two makeshift paddles.

So, there's me, the great boat builder. I got the two halves over beside my billet and did a little work on the canoe every day. Cutting both pieces so that they were about as true as possible, I brought them together with about a foot overlap, and did my best to make a good tight joint by lacing (riveting) them together. It certainly wasn't as nice a job as my stepfather could have done, but it didn't look too bad. I figured that somehow I would be able to get it to the Great Bitter Lake, just two or three miles away.

I could hardly believe it after the motorcycle theft, but a couple of Army blokes *stole the canoe*, took it to the Sweet Water Canal and hopped into it. However, I was vindicated when I found out what a *wonderfully lousy* repair job I had done! As soon as the thieves got into it, the canoes' halves parted company in the middle and promptly dumped both of them into the filthy water. Remember me saying that if anyone fell into that canal, they had to get something like seven injections? *How sweet (water) it was!* I only wished that the English corporal who ruined my bike could have joined them. That would have been even sweeter still!

~ * ~

Also while I was stationed at the Army camp, a new bloke (RAF) arrived in our billet. When we introduced ourselves, he told me that his name was Tom Roe and that he was from Motherwell. I told him that I had a pen-pal in Motherwell whose name was Mary Fraser and mentioned her address.

"Och aye, ah ken her fine, (Oh yes, I know her well) she steis (stays or lives - the "*ei*" sounds much the same as pronounced in "stein") just roon the corner frae ma hoose, but everyone kens her as 'Maisie.' She's a nice wee lassie."

When I next wrote to Mary, I was able to tell her that an old friend of hers (actually her brother's friend; his sister Marjory was *her* friend) was stationed beside me. Tom was a really nice chap and later I was saddened to learn that he had died just before I was released from the RAF. He couldn't have been any more than 23 or so!

Life was reasonably quiet from then on. I kept going into the Administration Office and asking for a transfer back to the Air Force, but not too much happened in that direction for some time. Then one day when I got off duty, I was sent for. "Your transfer has come through; you are returning to the Air Force immediately." Wonderful, and hard to believe! I rushed to my tent and started to throw everything into my kit bag.

Now, over the course of time, I had managed to secure a big, steel wardrobe for myself. In my haste to empty it, I filled my kit bag just about as fast as I could in case the Army orders changed. I had emptied the wardrobe, turned away with my kit bag, and suddenly thought, "Check again." In my haste, I opened the two steel doors, shoved my head inside to make sure that I had emptied it, saw "nothing there" and immediately shut the two doors. My head was still inside! I nearly knocked myself out and my backside landed flat on the floor as my head went round and round!

~ * ~

Two Egyptian lads pose with me at al-Fayid, 1953.

It took the Land Rover a little less than half an hour to take me to RAF Fayid, a few miles from the Arab village of Al Fayid, (where I was looking for the corkscrew) and just about the centre of the Canal Zone. (Somehow most of the Arab villages started with "Al." The Arabic name for the village at Abu-Suer was Al Abu-Suer.) There wasn't a band at RAF Fayid. I can't remember exactly what types of aircraft were there, but most likely Lancasters or Yorks.

~ * ~

On Christmas Eve, 1953 (6:00 p.m.) I had to report for "Fire Picket." Just prior to going on duty, the corporal who was in charge lined us up outside the fire picket building. He warned us that, as it was Christmas Eve, if anyone got drunk he would be put on a charge. Then we were dismissed and told to go to our billets and stay there until we were needed (if there was a fire and if the siren sounded). I went back to my billet, relaxed a little and one of the blokes gave me a drink, saying, "Here, have a little one on us." I don't know what it was but it was nice, so I had another…and another…

The fire alarm bell *apparently* went off somewhere around 1:30 a.m. as there was a fire in the NAFFI (canteen). Unfortunately, I was "blotto"—totally out of it! I can remember being in the ablutions (washrooms) as sick as a dog the next morning, and then having to take myself to the fire picket and report for duty. Everything was in a rather queer ethereal glow of light. I certainly wasn't myself. I can remember walking (more like "floating" slightly *above* the ground) towards the fire picket building. All the lads there stood outside in the brilliant sunshine, pointing and laughing their heads off at me as I staggered towards them. And you know what? I *didn't* care! I was still slightly sloshed!

"Okay, you people…look at this horrible individual…dragging himself here. What do you think of him?" the corporal shouted. "What do you think of him?" (I knew, roughly, that I was in trouble.) The corporal continued, "This is what happens when you are *man enough* to make the best of things…to enjoy yourself at Christmas time."

I was excused! How lucky can you get? I knew I wasn't "man enough" to enjoy myself but I certainly wasn't going to tell the corporal that—the lovely idiot that he was!

~ * ~

My first two friends at that RAF camp were an Irishman and an Englishman. Jimmy Newlands came from Belfast; the other was called John Aloysius Jones. One night shortly after I had gone to bed the two of them came to my bedside to tell me that John's mother had just died, and asked if I could lend him five pounds to get him home. I think that was all I had at the time but I gave it to him. Plus that, he didn't have any pyjamas and I ended up lending him a pair of mine. You guessed it! I never saw the money or the pyjamas again. I was screwed over once more by *another* Englishman! Never again, I thought! And how come it was always *five* pounds? Was that the *only* figure they knew? It would have been better for me if they'd known how to ask for two! That was when I made a vow that it didn't matter how sad a story I got from one of them, he would get nothing from me. He could find some other sucker. It seemed that it was always my misfortune to meet up with an airman from England. No, that's wrong; it was my misfortune to *lend* one money!

~ * ~

Shortly after this, an order went onto the daily order board that anyone who could play an instrument should report to the band room immediately. I was there as quick as a flash and picked out a cornet. I started to tootle and the person (Chief Technician Bull) who was in charge of the band came over and said to me that he was glad that I was in the band. Life from then on was a little more acceptable, as most of the time we were practicing to get the band into shape. We weren't great, but we made a fairly decent sound when the airmen went on parade.

There weren't many capable players. Alec Lumsden (remember his name) from Bathgate was my backup on cornet and a very nice player. I always thought that he was a better player than I was, but Alec insisted that he wasn't. It really didn't matter that much as we generally took turns at carrying the load in playing any solos that came along.

To get the band up to strength, I took to teaching a couple of my friends how to play. One bloke, Archie "Something-or-Other" from Edinburgh was a natural and caught on very quickly. I think it only took him about four weeks, and he was playing euphonium in the band; not great, mind you, but still quite a feat, indeed. Anyway, the band was formed and we were doing all right.

One incident that springs to mind was the time when four massive "Globe-trotters" had to land at our place to get refueled for the rest of their journey. (These were big American aircraft—they may have been C-130's or similar—which were en-route to the Far East for some reason that we weren't told about.) Apparently there was some kind of international rule that those on board were allowed to wear civilian clothing instead of uniforms, if a touch down was needed for refueling in a land foreign to the USA.

It seemed the whole station was to be on parade to welcome these Americans; plus that, the band had to have lots of practice so that we could show the Americans just how smart an outfit we were. So here we were, all nicely lined up and waiting. The first aircraft landed and taxied up beside us.

As the Station CO and his entourage marched up to just underneath the door of this aircraft, the door opened and an American appeared. Smoking a great big fat cigar, he was decked out in a baseball cap and a Hawaiian-style shirt displaying a naked lady leaning against a palm tree. This "vision" leaned against the side of the aircraft's rear door as the CO asked (in a "stiff upper lip" English accent), "Are you the captain of this aircraft?"

"Yep, ah guess ah am, Bud; ah guess ah am," was the reply. Now *that's* precious! It took us blokes in the band (who witnessed this) all our effort to control ourselves as we were on the verge of breaking up. I don't think our CO knew what to do or where to look; he was so stunned by this lack of formality!

After the welcoming bash was over, we dispersed to do our usual duties. Some of us had to help in refueling these monsters. The visitors stayed overnight and were to leave the next morning. A couple of us were talking to some Americans when the first pilot taxied to the end of the runway, revved his engines and let her rip. The Americans beside us gave him all the help he (the pilot) needed by shouting, "Give her more gas, Joe; give her more gas!" and whipping their arms as if that would help to get the aircraft off the ground!

~ * ~

Nobody slept indoors during the summer. The normal temperature during the day was around 110-degrees Fahrenheit (44 Celsius) with high humidity. To combat this a little, we would start work at 6:00 in the morning and finish around 2:30 in the afternoon. When we left the hangars to go for lunch at around 11:00, wearing our khaki drills (KDs for short—lightweight summer uniforms), we would purposely walk through the first showers we came to, soaking our clothes (keeping our shoes and socks on, too). By the time we got to our billets, picked up our plates, enamel mug and "irons" (knife, fork and spoon) and arrived at the cookhouse, our uniforms would be almost dry.

We also loved to sing as we "marched" along through the hangars and showers. We made up a jazzy version of the good old Scots' song, "My Bonnie Lies Over the Ocean" to which some of us would add "bum-puddy-dump-bump-bump" to add rhythm at the appropriate times.

(That also reminds me of some of the alternate titles our band gave to a few well-known tunes we played. For instance, Grieg's march, "In the Halls of the Mountain King" became "Dance of the Crabs on the Balls of the Mountain King." I also remember singing a song which ended with, "Singing Nellie, keep your belly close to mine!" Remember, we were deprived of female company for the most part, so had to have other interests to keep our minds occupied!

~ * ~

During this time of the year, almost everyone took their beds out to the verandah. The rule of thumb was that the bed had to go on the outside of the wall exactly opposite where it would have been on the inside. There were hooks at those exact spots from which we hung our mosquito nets.

One day after I had just come off duty, I noticed that a group of blokes were crowded around a bed on the verandah. I wondered what they were looking at and joined them, of course! They were in the process of following a line of bed bugs from the wall that was next to a bed, until it disappeared inside the bed's pillow and mattress. Someone said that the pillow should be opened to see how many were inside and we all agreed. A knife was produced and the pillow cut from one end to the other. There were *millions* of them—just one living mass! Unbelievable! It's a wonder we hadn't noticed this phenomenon earlier. (No one suggested opening the mattress.)

How on earth the airman who's bed this was, was able to hide this infestation for so long mystified all of us, for he was bound to have been well aware of the situation. The medics were sent for. They set about fumigating the whole place, from one end to the other. The bedclothes, pillow and mattress were taken out from the verandah, doused with petrol and set on fire. Someone suggested also doing that with the *airman* whose bed it was, for he was bound to have been aware of his "pets!"

~*~

There was another wee Scotsman there (from Glasgow), called Luke McFarlane, who was a casual friend of mine. One day just before pay day he asked me if I could lend him five pounds "until tomorrow and Ah'll give it right back to ye." Okay, stupid me—I never saw that fiver again either. You'd think I had a sign sticking out of my head saying, "Here's a sucker, ask him for five pounds!" Here's me blaming Englishmen! The one to blame was Ian for being such a sucker!

~ * ~

Around Easter, 1954, a notice appeared on the board that airmen, capable of qualifying, would be able to go to Nicosia, Cypress for a weekend, courtesy of the RAF. My buddy and I applied and a little while later we were told that we could go, and were given a date and time to fly there. It so happened that it was to occur on the same date as the next CO's parade. This didn't concern me; but it did concern the bandmaster. He wanted me to skip going to Cypress so that he would have me for the band. I said "definitely no way," that it would probably be the only time in my life that I would ever get the opportunity to see Nicosia and that I was going.

Because of this, the bandmaster *canceled* the band for the parade! You know what? Because the CO couldn't have a band, *he* canceled the parade. Boy, did I ever feel important! I had caused the CO's parade to be put off! Well, how's that? The lads in my billet all patted me on the back, others too, saying that it was great to have the parade "squashed!"

~ * ~

Well, I went to Cypress and found out that it wasn't that much different from anywhere else. I have a few photographs of some old buildings that were cracked by an earthquake a long time ago, and that's it. It really wouldn't have mattered one way or another if I had never gone. Well, at least I scrapped the CO's parade. It also meant that I had made a visit to *Asia*. I could than say that I had been on three continents: Europe, Africa and now Asia! (And later I would be in North America, and that would make four.) That only left Antarctica, Australia and South America for some later date! Hah!

I played solo coronet (closest to camera) with the
RAF Fayld band, Egypt, 1953/4. RAF photo.

It wasn't long after the Nicosia trip that the bandmaster approached me and asked me when I would be due to return to Britain. "Not long" I told him, "no more than two months; should be around the middle of June."

He told me that our band had been chosen to make a record to play for the Forces Radio Station which broadcasted all over the Canal Zone area. I told him that I thought that was a great compliment, especially as the band hadn't been going that long. What he really wanted was for me to sign on for another six months' service in the Canal Zone so that I would be there to make this record! I told him, "No," I wanted to go home—two and a half years out of my life were enough. He offered to give me an original disc of the recording if I would stay, but I still said "No." I don't know whether they made the recording or not; I really didn't care. I wanted to get back to Britain, away from the smell of the Suez Canal Zone.

~ * ~

Before I left Egypt, I got to know one of the pilots who often flew to the Far East. I asked him if he could get something for me to take home to Scotland that would be unusual. He said that he could get me a nice china tea-set that he thought I would like. I gave him the amount of money he said it would cost (can't remember how much). He brought back a beautiful set decorated with a dragon on the side of the cups, saucers, plates, and teapot—everything. When the cups were upended so that the light tried to shine through the bottom, the face of an oriental lady "magically" appeared in full colour. This type of fine china is called "eggshell." It was terrific, I thought, something special that I definitely couldn't buy in Scotland.

Here's me, shopping at an Arab bazaar, Egypt, 1952.

Other souvenirs I brought back from Egypt were two photo albums with camels and other desert scenes on the covers plus a number of photos that had been taken with my very first camera—a neat little Japanese number that was another "good buy" I had found in Egypt. It was a beautiful camera that had an expensive, fold-up tripod to go with it. The camera was excellent, with lots of "f" stops, speeds, and all the goodies that it took to take first-class photographs in those days. It also had a "delayed action" arrangement for setting it so that you could take your own photo. I bought this camera in 1953, so this technology was quite new.

~ * ~

I couldn't believe I was actually going *home*. It seemed as if it weren't quite real. All of the waiting for those last few months to pass seemed to take forever. A couple of months earlier, I had managed to secure a huge wooden box so that I could get extra belongings onto the ship that was to take us through the Mediterranean Sea. We boarded the troopship "Devonshire" at Port Said, at the north entrance of the canal and had to wait there for quite a while. (I don't know why, I know it was nothing to do with the tide, for the "Med" wasn't affected by tides. It's the only sea that isn't—I remembered that information from grammar school.)

This is where I saw what was called the "bumboats," probably because the small Arab boats were vying for trade (bumming) around the troopship, trying to sell nice handbags, purses and other types of goods to the men on board. An Arab would throw a rope up to the deck where it was secured; a customer would make his selection and a deal would be made, with the customer and the seller yelling back and forth. Then the Arab would put the article into a basket, it would be hauled up, and in turn the purchaser would put his money into the basket. This required a lot of trust on the part of the Arab; as it would be difficult, if not impossible, for him to get his money if the person refused to pay for the goods he had just purchased. There were quite a few of these boats around the ship, and the trade was brisk. I bought a nice handbag with

Egyptian hieroglyphics embossed on it (for my future wife as it turned out).

~ * ~

Then the classical *idiot* came along and an unfortunate thing happened. Around the sides of the troopship were heavy, round, iron "donuts," which were slid over upright metal rods. They were about three inches (7 cm.) thick by maybe seven inches (18 cm.) in diameter, with a hole in the middle for going over and onto the rod. (They probably had something to do with the ballast of the ship, for there were lots of them on each side.) Well, this idiot lifted one of them off the rod (he had to lift it almost three feet) and made out that he was going to throw it down at the little wooden boat about 20 feet (ca. 6 m.) below us. On the second movement of his hands, he let it go and it went straight through the bottom of the Arab's little wooden boat!

Lots of the blokes thought this was funny and had a big laugh at the poor, unfortunate man sinking in his little boat, trying to keep himself afloat and his stock of goods from being lost. The man could have been killed, either by the iron thing hitting him, or by drowning. I was upset that no one else seemed to consider that this poor soul probably had a family and found it hard enough to feed them as it was, and that this *mindless idiot* had just removed his livelihood.

No wonder the Arabs hated the English! There were times when talking to Arabs, I would tell them I was not English, that I was Scottish. The reply would be, "Ah, Escotia," and the Egyptian would pat me on the back, saying, "Quois, Effendi" meaning "Good, Sir." It seemed to me that the English of that era were branded similar to how the "ugly American" is branded today. You just had to look to India for proof of that! I didn't know that the man who threw the iron thing at the boat *was* English; just that he would be *branded* as one anyway.

* ~ * ~ *

A street photographer caught my first day out with my pen-pal, Mary, in Motherwell, Scotland, 1954.

CHAPTER SIX

Back to "Dear Old Blighty"

The troopship "Devonshire" got underway for the long journey back to Britain or "dear old Blighty," as we liked to call her. (This term was traced back to the British Raj in India. It was especially popular in the First World War where soldiers expressed a longing to return to "dear old Blighty" rather than fighting in the trenches. British music halls in the early part of the 20th Century used the term in such songs as "Take Me Back to Dear Old Blighty.")

As we got ourselves established on board, we had to report to a central office. This is where they compiled a work rota. Each person was given a berth for sleeping and a work number so that everybody did a little job of some sort during the long voyage.

Word went around that there were tons of *butter* in the galley, and good bread too; so I headed down there, only to find a whole bunch of other lads there before me. The cook was only too happy to let us help ourselves, although he wanted to know why we were feasting on bread and margarine. "No, this is butter, isn't it?" someone asked. "No, it's Stork margarine" was the reply. Boy, did we feel daft! If I had known then what I know now, I would have made some money from the company that made that margarine, to sell the advertising story of how it fooled us poor souls who hadn't tasted butter for years. I think the company's slogan was that you couldn't tell it from butter.

~ * ~

The trip through the Mediterranean Sea, from Port Said to Gibraltar, would have been about 2300 miles (3700 km.) and the troopship, at about 22 knots' speed, (I'm guessing) would take around four and a half days. Anyway, I was looking forward so much to seeing the Strait of Gibraltar (called the "Pillars of Hercules" in ancient times), as it is just a short distance between the land points of Europe and Africa, maybe around 15 miles (24 km.) and there would be lots to see. No such luck! It was the middle of the night when we passed through. Then we sailed up the coast of Portugal (visible in the distance) to the Bay of Biscay and all the way up the coast of England to Liverpool. (Thinking back, that trip must have been further than the voyage I later took from Liverpool to Canada—close to 4000 miles.)

~ * ~

I don't know what stage we were at in the journey when I was talking to some commandos on the ship and happened to mention that I was from Campbeltown. They informed me that there was a Campbeltown man in their group. I asked what his name was and they told me. It didn't mean much to me.

A day or so later I heard a voice behind me shouting, "Ian—Ian Morrans!" I turned around to see who it was that knew my name. This big fellow was shaking my hand and I wanted to know who he was. He told me he was Hector McMillan and I didn't recognize the name.

"Well, ye'll remember 'Pinhead,'" he said and I said "Aye, Ah do."

He replied, "Ah'm Pinhead." I didn't want to believe it, because we used to beat up Pinhead for fun when I was a boy … and this bloke was about six or seven inches taller than I was. (It would have been easy to be taller than me, for I was under 5 feet five at the time—ending up even shorter as I approached my later years.) Plus that, he was a *commando* to boot! We talked for some time and thought about going for a beer. I told him that you would have to be desperate to drink the beer on board, as it was the worst in the world—tasteless and all full of sediment from continuously being shaken by the motion of the ship.

~ * ~

I had been writing to Mary Fraser, my pen-pal in Motherwell, for probably about a year and a half while I was in Egypt. When I knew I was going "home," I had written to Mary and arranged to meet her *under the clock* in Central Station in Glasgow on a certain date (time to be arranged). This was the only landmark I knew in Glasgow. (Later I found out that that was *not* a good place to arrange a meeting, as it was well known, apparently, as a place for meeting prostitutes. I, of course, didn't know this, not being from Glasgow; but Mary braved the thought that someone might proposition her, and waited there for me just the same.)

We onboard the Devonshire had already been informed of our arrival date in Liverpool but didn't know the time of docking or the time schedule for the train to Glasgow. As soon as we were settled into the waiting area at the railway station in Liverpool, I made enquiries and found out. I then sent a telegram to let Mary know the time of my arrival in Glasgow.

The train trip went well—east to Warrington, north to Preston, Lancaster, Penrith, Carlisle and finally Glasgow. (Little did I know that I, with future family in tow, would be taking the exact same route back down in a few years' time but, instead, would then be sailing west over the Atlantic.) Luke McFarlane was in the same compartment as I was!

"Okay, Luke, where's m' five pounds?"

"Ah've nae money on me, Ian; ah wiz wan o' them that hudnae got peyed. (I was one of the people who *hadn't got paid*). Tell me whaur ye stey an' Ah'll sen' it tae ye." (His

Glaswegian was even thicker than most!) It didn't happen. (I've since forgiven (if not forgotten) those debts in my mind and decided *none* of the people involved in this episode of my life owe me *anything*.)

~ * ~

Mary was there, waiting for me. I knew her right away as we had previously exchanged photos. (And you know what? She was the *only* female standing under the clock!) I can't remember whether I kissed her, gave her a hug or just shook hands with her—most likely the latter. Mary was very shy, so I can't imagine her allowing anyone to kiss her in *broad* daylight, not even a wee peck. (Goodness gracious me, that would *never* do! That would be something akin to making love in the middle of the street in broad daylight in *her* mind. It wouldn't have been any more acceptable to make love in the middle of the street *after dark*, either, come to think of it. She did ease up considerably over the years, though; thank goodness!)

~ * ~

We got a bus from Glasgow to the nearby town of Motherwell, then a local bus to Mary's home to meet her mother, Susan Fraser. Her mother welcomed me as if she had known me for years and told me that I could spend as much of my leave at her house as I wished. We had something to eat, and later Mary asked me if I would like to go and meet her Aunt Jean and Uncle Jimmy in Bothwell, a suburb of Glasgow. I said I didn't mind. It turned out that they were the parents of Billy Russell who gave me Mary's address (in Egypt) so that we could be pen-pals.

On the way there, Mary told me not to pay too much attention to the way her uncle spoke, as he was "actually really very nice." I didn't pay a lot of heed to what she said, thinking that, well maybe he had a defect in his speech which made him difficult to understand. It was nothing of the sort. I discovered rather quickly that Uncle Jimmy had a very colourful vocabulary that could shame the Devil himself!

Mary and I got off the bus in Bothwell and were walking towards their house when who should we meet but Uncle Jimmy, heading for the pub. When we were introduced, he held out his hand and said, "Ian, well, well, Ah've waited two f—ing years tae meet ye." I was flabbergasted! Nobody spoke like that in public in those days (and I still don't like it or use it today even though we're a lot more liberal today than then. To me it indicates an inability to express oneself using proper English, and it *is* true if you think about it.) Anyway, as soon as he opened his mouth I realized what Mary had meant by not paying too much attention to the way he spoke.

He told me that he had heard a lot about me from Billy and that he was really pleased to finally meet me. He certainly was a very nice person. We had many a good game of cards in the ensuing years. At first I found it difficult to listen to him swearing all the time, but I soon got used to it. Of course, I was used to it in the Air Force, but it wasn't usual to hear swearing amongst mixed company. Rough language like that was *not* used by females, either. If a female was heard using bad language, she would be shunned like a leper! (I thought it might have been picked up by our two very young daughters later, as we did quite a lot of visiting over the years. But not once did either of our girls repeat what Uncle Jimmy was saying when he spoke.)

Billy was at the house and it was nice to see him again. I brought him up to date on all that had happened since we had last seen each other.

~ * ~

I had a month's disembarkation leave and spent most of it at Mary's house, with a few days down at Campbeltown to spend some time with my mother and Bill. During my stay in Egypt, they had been given a new house by the local council. Their new address included a totally new street name, in a part of the town that hadn't existed before I had left. So on my first visit home, I had to find out where *I* lived! This completely new area was all filled with different types of new-style homes. Theirs was called a "prefab" (previously fabricated) house, meaning that it was built in preformed sections and assembled at the site. It had a life-span of 15 years, but was stretched out to more than 20 years. It was really a nice little two bedroom house with about 700 square feet in area. (Permanent housing now occupies that area, but it was a means of "sudden" housing at a time of need.) I was really pleased at how pleasant it all was.

We had a good reunion and got caught up with all the local news. I asked Mother what had happened to Lorne Street and she told me that it really didn't exist anymore for housing. All the buildings had been torn down and what was now a solid wall at the edge of the sidewalk had become the rear wall of McConnachie's garage. There was no row boat to take fishing. I figured Bill must have sold it, so I didn't ask.

~ * ~

My large wooden crate from Egypt had been addressed to Mary's house. (I must have had confidence that everything would turn out okay there.) It arrived shortly after I did. The first thing I checked on was the eggshell china. I had originally received it in a small, wooden box, all nicely wrapped and packaged, and I kept it that way. It was in good shape.

I think I had only got to know Mary in person for about three weeks when I asked her if she would marry me. She said, "Yes." This was about the middle of July 1954. I gave her the eggshell china as a present. I had previously given her the nice Egyptian handbag I had purchased from the "bum boat." She proudly displayed it everywhere we went. Then I thought that I should buy an engagement ring, but I would have to find more money to do that.

I still had a year to go in the RAF and we decided that we should get married the coming December. We had agreed to pool our money to pay for the wedding as Mary's mother was a widow and probably not much better off than my mother was.

~ * ~

Because I didn't have enough money to buy an engagement ring for Mary, I reluctantly decided that I needed to sell my prize possessions from Egypt—the camera and tripod which were also packed in the big wooden box.

I headed into Glasgow with them. The only area that I knew there was the Saltmarket (near the intersection of Argyll Street and High Street and not too far from Steele Street where we used to shoot the rats when I was a lot younger). I thought there were quite a few shops in that area that might be interested in what I had to sell; plus that, it was convenient as the bus route was quite close.

There were a couple of things against me right from the start. One was that I didn't have any civilian clothes to wear, only my Air Force uniform. That indicated right away that, as I was in the Forces and walking into their shop with something to sell, it was likely that I didn't have very much money. Next was the fact that my accent (from the Western Highlands and quite different from Glaswegian), gave the shopkeeper (and others too) the impression that they were dealing with a "country bumpkin" who didn't know too much. (I guess I have to admit this was partly true at the time!)

The owner of the shop went through the rigmarole that he was overloaded with cameras and didn't really need mine. This was impressed upon me by a wave of his hand to indicate the cameras he had in the shop window. But, he would "take the things off yer hands" to help me, and he would take a chance and give me five pounds ($15) for them! If I had known then what I know today (how often have I said that?), I would have walked out and gone to other places to shop around. Anyway,

I took the five pounds, added it to the money I had and got Mary's ring. It all boils down to experience (or lack of it). If I'd had time later, I would have loved to go back to that shop to see what price tags he had put on the camera and tripod. W-e-e-ll, maybe I was better off not knowing. I'm proud to admit now that I have learned to barter with the best of them. I can usually get a used object for my price, or at least at the lowest possible price. No more the "country bumpkin!"

~ * ~

Before I had disembarked at Liverpool, I had been issued with all my back leave (a month's worth) and informed that my new posting was in Scotland. I was going to RAF Kinloss. I had to report there at the end of my leave. This was the station I flew to for that unfortunate weekend that got me in trouble and docked me two days' pay away back in 1951!

Kinloss Royal Air Force Station was quite northerly, about 28 miles (45 km.) east of Inverness. It was close to the little town of Forres, in between Nairn and Elgin. As I went round with my "arrival form," of course, the first thing I asked about was if there was a band. Oh goodie, there was! I wasn't long in settling in and got into the band right away. I was also asked if I would play for the dance band as their last trumpet player had just been returned to civilian life. The dance band, a five-piece combo, consisted of tenor sax, alto sax, piano, drums and trumpet (me). We also had a singer, Jim Nelson, from Glasgow. The two of us got along really well. His bed was in the corner and my "bed-space" was right next to his. When I was "at home" in Motherwell for weekends, Mary's mother always did a lot of baking for me to take back to camp. Jim and I would have a feast when we got into camp late at night after playing at a dance.

We must have made a decent sound, as we, quite often, were invited to different nearby towns to play at their local dances—such places as Elgin, Nairn, Forres, Huntly and other small towns. The most memorable, by far, was the invitation we got from the small town of Conan, near Strathpeffer, a little west of Inverness. We were treated royally—far better there than anywhere else we ever went. The town council put on a welcoming committee for us. There were a couple of speeches and a meal was laid on. You would have thought a big name dance orchestra was visiting the town, not a "two-bit" outfit like ours!

~ * ~

Most of the dances we played for, however, were at the camp, sometimes on a Friday and sometimes on a Saturday, except for one Saturday each month. That was when everyone was allowed to go home for the weekend. I think the first or second weekend I was there was the free weekend. I was told that there were lots of busses laid on for the men, some going to Glasgow and others going to Edinburgh, but that I had to get there early, as it was often hard to find a decent seat.

I thought I was being fairly early in getting to the busses, but I soon found out that wasn't the case. All the Glasgow busses were lined up, and as I looked in each one I found it full. I kept looking, and there, in the middle of the third or fourth bus was one empty seat. Boy, was I lucky! (Or was I?)

I guess the blokes who were on the bus had a good chuckle when I took the seat, for I didn't know that I was sitting next to the "village idiot!" And I had to suffer him *all the way to Glasgow*, as no one got off the bus in between. From the time the bus started off, this idiot ranted to me about this and that. It didn't take long for me to start praying that he'd lose his voice. He went on and on. I even tried to make believe that I was going to sleep. That didn't work. He would shake my arm to make sure I was awake and paying full attention to what he was on about. You wouldn't believe the things he was telling me—how his mother always locked him in a closet when he was a little boy to make him behave—lots of things like that. My ears had practically burned off by the time we got to Glasgow! I also realized why that seat had been empty. All of the other blokes preferred to let the idiot sit by himself!!

I was more fortunate on the return journey. I did ask about "the idiot" later and was told that everybody steered clear of him. Some said that he was trying to "work his ticket," meaning that he was only acting this way all the time so that he would get discharged. Others said, "No, that bloke is really gone."

There were various stories going around about this chap and I will relate a couple that happened just shortly after my encounter with him. One time he marched into the Wing Commander's office, straight past the clerk, then sat down and looked at the officer.

"*What do you think you are doing?*" the officer yelled. Without blinking, the idiot stated that he had come for his tea and that he had been invited. (You have to think about this one, for if he really was "working his ticket" then he had to know for sure that the officer was in his office, to achieve the greatest effect. No good walking into an empty office.)

There was a saying that was going about at that time. If someone got a little peeved with you, for example, or maybe you were annoying them a little, they might say, "Oh, why don't you go home for two weeks."

Well, a corporal friend of mine was working on an aircraft engine one day, and apparently this nutty guy was annoying him constantly. The corporal turned round, and, without thinking, said to him, "Go away; go home for a month." This could maybe be another measure of the bloke's real intelligence too, for he *did go home*, which is where he wanted to be. When the RAF police finally found him at his home, he replied that he only did as the corporal had ordered! I never did hear the end result, nor did I ever see him again. (It was a fairly large camp, and I only *did* see him in person that once, so maybe he was still there or maybe he eventually achieved an early discharge.)

~ * ~

One time a top-secret American aircraft landed and was taken into a hangar. I went to see if I could possibly have a look at it. I had been told there were guards at the door but didn't find any, so I just strolled inside. When I asked an American at a table nearby if it was all right for me to have a look at this aircraft, he replied, "Look all ya like, Buddy. Ah's jes da cook."

~ * ~

We spent Tuesday evenings crowded around the radio. That was when "The Red Planet" was on, a sci-fi story about Mars, followed by "The Flying Doctor," about a physician in Australia. (We had to use our imagination in those days, with the pictures in our heads; not like it is today with television.)

As in Egypt, it seemed as if I was the rare one who could make his money stretch. A friend at Kinloss started coming to me on a Monday, asking to borrow 10 shillings. At first that was okay for he was always prompt at repaying it on pay day. The only thing was that this went on week after week until one time he came to me to repay his debt and I told him to keep it. He didn't understand why I said this. I replied that I didn't know whose 10 shillings it was anymore, so he'd better keep it and not come back. I explained to him that now that he had an extra 10 shillings in his pocket, he wouldn't have to borrow from anyone anymore. The borrowing stopped.

~ * ~

When it got near to our wedding date, I had to borrow a friend's motorcycle to make a trip into Elgin to buy a wedding ring for Mary. I suddenly realized that I would be going "home" that coming weekend to get married, so I had to go and get the ring, like *right then*!

There was no way of getting to Elgin through the week as no busses ran there. The only way was to bluff it. I got into my dress uniform like I was on leave or at least had permission to leave camp, borrowed the motorcycle, drove up to the guardhouse, stopped and gave them a wave, then drove out the gate. I went into Elgin, got the ring and returned to camp, giving the guardroom blokes another wave as I entered. Whew, I had made it! By the way, I *didn't even have a driving license* at the time!

~ * ~

Mary and I were married on December 29, 1954.

Mary and I were married in Craigneuk Parish Church in the Burgh of Motherwell and Wishaw on the 29th day of December 1954. The bride wore white and I must say she was 100 per cent entitled to do so. This was not like today's brides, who wear white and it doesn't mean a thing—even if they've been married 20 times before! When we got married, white meant that the bride was pure, and if you want me to spell it out, it meant that my Mary was a virgin and thus *entitled* to wear white.

We had a three-day honeymoon in Glasgow at Mary's Uncle David and Aunt Liz's house as they were not going to be there for a few days. Then it was back to Mary's mother's house for a day, and then I returned to Kinloss RAF.

I think it was during a couple days' leave for Easter break when I went to an auction market in Hamilton (the next town, about 3 miles or 5 km. west of Motherwell) one Monday afternoon. There I bought an old piano. I paid ten shillings for it as I was the only one bidding. Even I wouldn't have normally bothered with it, but when it got down to that ridiculously low amount, I put my hand up. No one else wanted it! This was about a dollar and a half ... *for a piano*! One foot was missing, which caused it to sit at an angle. The ivory was gone from a couple of the white keys but it was in perfect tune, sounding and working very well.

I was at the dinner table, getting ready to eat when Mary got home from work. During our meal she asked me what kind of day I'd had. Well, I told her that I had gone to the auction market and bought a piano. She asked me how much I paid for it as we didn't have money for anything like that. I told her. Well, she laughed so much that the tears were rolling down her cheeks. Then suddenly she changed and said, "Ye're no bringin' an old piano like that into ma mother's hoose!"

It didn't seem to matter how much I told her that it was not such a bad piano, she maintained that any piano that could be bought for that amount of money was only good for firewood. We were all laughing during this dialogue as it did really seem very funny. I agreed that I should go back to the auction place the next day and tell them to put it up for sale as I didn't want it. I gave them my Air Force address and a couple of weeks later I got a cheque for fourteen pounds, ($42) and that was after they had taken off their commission. So I really did all right out of that deal! I think that was the week I told the pesky borrower to keep the 10 shillings, as I was probably feeling "rich."

Towards the middle of the summer after about a half year commuting to be with my bride every second weekend, I acted on Mary's idea of us renting a room at a house in Forres near my RAF station. Making some enquiries, I found a room that we could have for a week, got a date arranged and Mary came up. I got her to the little house where we had rented the room and the lady told us that we could also have the use of the kitchen so that we could make our own meals. Moreover, I had received permission to remain out of camp at night for the length of her visit. (My daily rate of pay was increased during that week, as I was then classed as "living out.")

Some time before this, I had taken Mary's brother's bicycle to the Air Force station to help me keep fit, as it was not used at all at home. A friend from Dundee and I used to go for bike rides during the evenings. This allowed me to make the trip from camp to Forres after duty and then back to camp in the morning, as it was only a few miles.

We had to buy groceries. (Actually, we in Scotland called it "goin' fer messages." The idea was that the housewife wrote out a list of items needed (messages), took them to the person behind the grocery counter, then the "messages" were filled by the clerk, packaged and then the customer paid. No such things as supermarkets back then.) Before this, Mary wouldn't eat butter—said she didn't like the taste of it. Now, when we were shopping at a local small grocery, she asked for butter. I asked why she was buying butter and she said that she didn't want to appear cheap to the lady of the house by getting margarine. I guess she must have realized just what she had been missing all these years, for she continued to buy butter from then on.

~ * ~

The following Saturday night, our dance band was playing at Forres Town Hall. I had been bragging to Mary about the introduction to the song, "Cherry Pink and Apple Blossom White," that I played on trumpet, standing up, before the rest of the band joined in. It consisted of the first three notes, then into a big *glissando*, using the third valve slowly, to go *down* and then *up* to the third note again and then continuing right into the melody when the rest of the band joined me. It was just a copy of what a big-time trumpet player (Maynard Ferguson) of that era did. Everyone thought it was very effective, sounding and looking quite professional. So, there's me saying, "Wait'll ye hear *me* play!"

Saturday night came and my Mary was sitting at the side of the hall, close to the band, her eyes firmly fixed on "Lover-Boy." Then it was time for me to shine. I stood up, the first two notes came out correctly, but I have no idea what happened to the bit where I was supposed to do the fancy stuff. I played absolutely terrible! The rest of the band started all right, but I had to sit down with a very red face—even redder than usual! In front of my Mary, too! You know, I must have played that "intro" at least 40 times previously without fail. (That's what I get for trying to show off!)

As this happened in the middle of the summer, I had only a couple more months to do before my release from the Air Force. The time passed quite quickly, and then I was "demobbed" (short for demobilized). And I also got my "demob" suit. This was a suit of civilian clothes that was given to everyone who was leaving the Forces, whether it was Army, Navy or Air Force. It was always such a horrible fit that everyone in sight knew what it was. Anyway, it was better than nothing, and I picked a navy-blue suit with a pinstripe. I didn't have it very long when the word went around that pinstripes were out of style! Oh well, I didn't like it anyway.

~ * ~

I hadn't used up much of my leave except for the couple of days I had tagged onto my break for the wedding. The way it usually worked was that the English airmen, who all preferred

Christmas, got *that* break and we Scots carried on our normal duties over the holiday. Then we got the break for Hogmanay (New Year's Eve—the biggest Scottish holiday) as this was generally the time preferred by us. I had added a couple of days before it and a day after it, saving most of my leave for when I finally got out. Mary and I figured that it would give me more time to get a job in case there was a shortage of work at the time. It also meant that I would get out earlier as my leave had to be used up by the time my finish date arrived.

~ * ~

Nothing to fear. I got a job right away at Redlac Engineering in Bellshill (pronounced "bells hill") as a machine-fitter building printing presses. It took about fifteen minutes to get there and it wasn't a bad job although the money was lousy. This I had no knowledge of, for I had no idea what the wages were in "civvy street," as we called it. I quite enjoyed working there and I probably would have stayed there longer if the wages had been a wee bit better. (I was getting four shillings an hour—around 60 cents in 1955.)

Around this time I learned that Mary had a certain trait I didn't like. (It seemed to me to be a "Fraser trait" for her sister Nettie had it, too, but even worse!) Mary and I didn't argue much during the first months of our marriage. (Actually, we didn't argue much either after I was home and working at a civilian job full time.) Then it so happened that every time Mary and I had a little tiff, she would stop talking to me for up to three weeks at a time, until it suited her to talk to me again! If I tried to start a conversation with her, she would ignore me. I suffered this for I don't know how long, maybe a couple of years, until one time we had a little falling out and she started with the silent treatment. At that point I thought, 'enough is enough.'

I approached her and said, "Mary, Ah'm makin' ye a promise right now, and Ah'm sincere about this. The next time Ah speak to ye and ye don't answer me, Ah will no speak tae ye again—ever. It will be yer choice. If ye ever want us tae start speakin' tae each other again, YE are the one who will ha'e tae break the silence—and Ah really mean what Ah'm sayin'." It took her about three days to come up to me and started talking to me again. That was her "carry-on" over for good as far as I was concerned. (I understand from my daughters that she used the silent treatment on them and their husbands over the years, though.)

~ * ~

After about six months, I got a job as an industrial mechanic at Carfin Leather Factory. This was all right for a while. I was the only person in that position and was supposed to repair broken machines and also keep the machines in good working condition so that they wouldn't break down and cause loss of production. In the trade, this was called "preventive maintenance." The factory was owned by two Jewish brothers, Mr. Hans and Mr. Ernest. Hans was all right. I would wash his car for him when I didn't have any repairs to see to and he would give me two shillings or a half-crown (two shillings and sixpence) for doing it. If I washed his brother's car, *he* (Mr. Ernest) would give me nothing, saying that, as I was employed by the company and as I was doing it on company time, I didn't deserve anything! He did have a point, for there was no union, so I couldn't complain to anyone. I made up for it by stretching out the overtime I would get some weekends, so that the repair I was working on was completed for starting first thing Monday morning.

This other (younger) brother, Mr. Ernest, would often complain to me that he didn't pay me to walk around all day doing nothing. And try as I might, I couldn't make him understand that I walked around checking on the machines, keeping them in good running condition. As soon as I started working, it meant that something was broken and he was losing production!

I must have told him this umpteen times in the time that I worked there, until I was sick of hearing it from him.

~ * ~

Deciding to look for another job, I found one at the Lanarkshire Welding Company in Craigneuk (the area where I got married). I was taken on at a bit more money than I had previously made. After I'd been there just a week, one of the overhead crane operators quit and I was asked if I would care to learn to drive the crane as there was quite a bit of overtime involved, Tuesday and Thursday evening 'till eight and all day Sunday—no Saturday. Hey, I needed money and this was just dandy! I said, "Sure, I would like to learn to drive the crane."

There was quite a skill to driving the crane up the shop, lowering the hook and traversing at the same time to arrive at the planned position quickly. Everything had to be right there just ready to hook on. It was the sort of job where one had to be good at judging time, distances, speed and position. Besides, I enjoyed being up in the air and watching all the guys working beneath me.

During a tea break (no coffee breaks for Brits), a few of us were discussing the Americans having put a rocket into space and the Russians with their Sputnik a few years earlier. I remember mentioning that it wouldn't surprise me to see a man on the moon within 20 years. Well, I got laughed at by everyone who was there. They thought it was funny that anyone could possibly think there could ever be a man on the moon. (It would have been nice to see all of their faces in 1969! I wonder what kind of reaction I would have got if I had said *10* years! They would probably have had the men in white coats come and take me away!)

~ * ~

Sometime after we were married, Mary and I were on holiday in Dunoon, Scotland. It so happened that this was where my good-looking cousin, Rose Morans (spelled a bit different than my last name), was working as a nurse in Dunoon General Hospital. I thought it would be nice if Mary could meet her and we could spend a few hours with her as I hadn't seen Rose for years. Mary and I went to the hospital and I asked for Rose at the reception desk. The person I spoke to told me to wait and she would get Rose for me. We waited for only a few minutes when I saw Rose literally flying down the staircase. She ran right up to me, threw her arms around me and gave me the biggest hug I'd had in a long time! I don't think Mary liked Rose too much after that.

We spoke for a few minutes and arranged to meet Rose that evening so we could go to a movie together. Well, when it came time to leave to meet Rose, Mary said, "We should stei hame, never mind aboot meetin' Rose."

I replied that we had made arrangements to go to the movies and that Rose would be waiting for us. Stubbornly, Mary refused to go so I countered, "Then Ah'll go by mysel'."

When I met up with Rose, she asked me where Mary was and I responded that she had decided not to go to the movie, trying not to make a big deal out of it. We headed for the cinema and when we were inside I decided to have a look around before the lights were lowered. Low and behold, I spotted *Mary* sitting about three rows behind us! There she was keeping tabs on me, and probably checking to see if Rose and I did any cuddling while we were in the cinema. She looked straight at me and I looked straight at her. Not a word was spoken; nor was anything mentioned about it when I got home after I said "goodnight" to Rose. (I sure wasn't going to bring it up! I've often wondered if Mary followed us when I was walking Rose back to the hospital after we left the cinema. That's a Fraser for you!)

When I first got out of the Air Force, Mary and I lived with her mother, who was in a two-storey, three-bedroom council house in Motherwell. (Council houses were owned by the local township and rent was paid monthly to the local council. The rent was lower than it would have been if the house were owned privately.) It was quite a nice house, with plenty of room. What complicated things, though, was that Mary's brother Charlie and his wife also lived there. Charlie's wife and Mary's mother didn't get along at all. His wife used the kitchen first to make their meals and then it was our turn. I didn't like this arrangement and told Mary that I was going to look for a place for us to rent.

We got a "single end" in Dalziel Street. ("Dalziel" is a Gaelic name, pronounced "dee-ell.") This was one room, old-style and similar to what I was raised in when I was a wee boy in Campbeltown. (Boy, I'd come a *long way* in twenty years!)

Well, there were a few improvements. At least it was a little bigger, had a real window, was on the ground floor, had an inside faucet with cold water, a decent white sink and a gas "geyser" to supply instant hot water at the sink. This was a small appliance that had a system of copper coils inside a sheet metal cylindrical casing that was about eight inches (20 cm.) in diameter by maybe 16 inches (40 cm.) high and was situated immediately above and to the side of the cold faucet. The water was heated by a gas jet. It wasn't great, but it was very handy when it came time to wash the dishes, your hair, yourself, whatever!

We did have an outside toilet, but one major difference was that we had electric light. I was 23 years of age, and for the first time in my life I could switch on my own electric light bulb. (Remember, I had come a long way!)

Still, it was home, and the two of us were happy there. Then came the big news. Mary was pregnant! This must have been discovered in January or February 1956, for our first child arrived in August 1956. It could maybe be interpreted as me giving Mary an early Christmas present in November 1955—count it out!!!

~ * ~

After our move, Mary's mother was continually at us to move to her house, as Charlie and his wife had finally moved out and she found the house too big for her. I had kept saying that we were fine as we were, although I did admit to myself that her suggestion made sense. I kept telling her that I would think about it and the final decision was made when she assured me that I could treat the place as my own, and that anything I wanted to do would be fine with her. It also made sense to make the move with a baby on the way.

Mary's mother was "Mum" (Mom over here) to me and we got on really well. Any time Mary and I had a disagreement; Mum would say, "Ian's right, Mary." I let this go on even though I knew that she really shouldn't have said so.

The house had a front lawn and a chain-link-fenced back yard. There were a few roses across the front of the house and a few more along the side. Thinking we needed a nice place to sit, I decided to put in a rose garden in the back yard with concrete winding paths. Then I constructed a garden bench that would seat four people, and set about transplanting the roses.

At that point Mum came out and asked me what I thought I was doing with *her* roses. I told her that she said that I could treat the place as my own and she replied, "Aye, anything but ma roses." What could I say to that? So I told her to keep her roses in the front and then I quit working—the side roses had already been transplanted so we had something to look at.

~ * ~

Sometime later, I got to studying a little about fiberglass bodies for cars. I figured that, if I could buy a half-decent chassis, then I could make a body around it and we could have a car. It was a crazy idea, but nevertheless I was engrossed in it, dreaming what sort of fancy body I would build to be the envy of the neighbourhood. Well, the old saying holds true, "If you're going to dream, it is just as cheap to dream big." So I let it slip one night that I planned to do this and here's what I got—"Ye're no puttin' an old wreck of a car in ma gairden!" So I stopped dreaming and doing, figuring that I had made a mistake by moving there, and now there was nothing I could do to fix it. Despite this, Mary's mother and I still got on well. Little differences were soon patched up and life went on.

~ * ~

All three of us agreed that the upstairs bathroom floor was looking a bit shabby and that we should buy floor tiles. Before Mary and her mother left to shop for them, I told Mary to make sure that the tiles she bought were "inlaid." There were two types of tiles on the market at the time. The cheaper tile had patterns printed thinly on top of the surface which wore off rather fast. The type that had the pattern inlaid right into the tile was more expensive but lasted forever.

When they arrived home again, Mary asked me if I was trying to make a fool of her. I asked her what she meant and she said that she asked the shop assistant if the tiles were inlaid and he replied "No, ye lay 'em yoursel'." So then I had the job of convincing her that it was the bloke in the shop that didn't know what he was talking about, that the actual name for them really was "inlaid tiles."

We had bought some cream coloured tiles and some greenish ones in equal numbers. I took them upstairs and laid them down loosely in a couple of lines on the bathroom floor to see how they would look, alternating the colours. When I had done this, I called for Mary to come up to get her opinion. She looked and said they were nice, but wished that she had got the *diamond-shaped* ones. I picked some up and rotated them forty-five degrees and put them back down again to let her see that there really *weren't* diamond-shaped ones; they just appeared to be when laid differently. She was a little embarrassed by her apparent stupidity!

~*~

Monday, August 13th, 1956, dawned and Mary was in Motherwell Maternity Hospital. I had the hard job of pacing the floor at home. As she had been admitted on the Saturday before, this meant that we had been playing the waiting game all weekend. The situation was very much different in those days in respect to the prospective father being with his wife in the delivery room. It just wasn't allowed. There was a pay phone about a quarter of a mile (1/2 km) from the house and I was calling every half hour. Then we got the news around noon. Mary had given birth to a wee girl at 11:45 a.m.

I thought that was great and Mum was thrilled. Later I had to go to the Cooperative Grocers to buy a loaf of bread and a pound of sausages. On the way there (walking, of course!), I had to pass the house of Mary's Uncle Tom and Aunt Sarah. Mary's mother had asked me to pop in and tell them the news. I went to their door and when Aunt Sarah opened it, I flustered out the "messages," instead of the announcement by saying, "A loaf of bread and a pound of sausages, please." She thought that was really funny and it took me a few years to live that down! (At least I didn't order a new daughter at the Cooperative!)

~ * ~

I visited the hospital as often as I could but there were more rules and regulations than there are today. On the third day after the birth Mary showed me that she was still lying on the same sheet that she gave birth on, which was badly stained with blood. I went to find whoever was in charge and gave them a blast that most likely was heard miles away! The sheets were changed before I went home and I "signed them out" the next day to take them home to where they would be better looked after. (The time spent in hospital then was a lot more than it is today!)

Mary and our new daughter, Audrey Susan Fraser Morrans, arrived home on August 17, 1956. We didn't have a lot of money to spare, but we did have a little celebration. Mary had an older sister, Nettie ("short" for Janet—but not really—actually it's longer!), and apparently there had been some sort of squabble some years before and they had been out of contact. The first time I had met Nettie was at our wedding. Before that Mary told me that Nettie had not spoken to her for years—my recollection says it was about six years—and I think that she also cut Mary's mother off, too. I don't know what caused the rift but Mary told me that when Nettie moved out of the family home she moved in with their grandmother. I also don't know how it came about (probably Mum's or the grandmother's instigation) but Nettie did come to the wedding and served as Mary's maid of honour. That was the last we had any contact with her until Audrey was born. Then Nettie appeared to see the baby and was a constant visitor. We could hardly have any time by ourselves weekends—she was always there. As far as I know, no one ever discussed the past problem. Nettie just came back and everyone was friendly again. After Audrey's arrival Nettie and Mary became the best of friends. (I never could figure out the Frasers!!!!)

Four generations pose following Audrey's baptism, 1956. From left to right: baby's aunt Nettie (Janet) Fraser; baby's great-grandmother, Janet Luxton, holding baby Audrey Susan Fraser Morrans; baby's grandmother, Susan Fraser; and baby's mother, Mary Fraser Morrans.

Mary's mother had gone to visit friends at St. Andrew's, accompanied by Mary and Baby Audrey. The plan was that Mary was just going to spend the weekend and return, leaving the baby with her mother for the fortnight (two weeks) and then go back at the end of that time to help her mother home with the baby.

(Before I go on with the rest of the story I need to explain that certain friends of mine had been on vacation and had brought a cute little ornament for us which Mary didn't know about—a rolling pin form with a nice, funny little verse on it.) Anyway, Mary's first trip was to be on Sunday night on the last bus from Glasgow, getting off at the Motherwell Bridge, level crossing bus stop at half an hour past midnight.

There were two routes a person could walk to arrive at this bus stop. The most direct way via Watling Street was along a dark, lonely road; the other along Fort Street was just a little bit longer but was well lit. We agreed before Mary went away that I would go up the well-lit way to meet her (as there was no way she would travel along that dark road on her own if the bus happened to be early). I also told her that I would make sure that I was at least ten minutes early, so that I should be there before she arrived.

I took good care to get there early so that I would be in plenty of time to meet the bus. I arrived and waited and waited and waited. It was, by then, well after the scheduled time and no bus had come. A police car stopped. One of the policemen got out and asked me why I was hanging around at that time of night. I told him that I was waiting for my wife who was supposed to be on the final bus. He then told me that it had gone through some time earlier. So I decided to head for home, thinking that she had missed the bus entirely.

The bus had been *very* early but Mary had caught it. She was a little concerned that she would have to wait a long time at the lonely bus stop for me to get there. Two other people had got off at the same time as she did and they started to head down the dark road, so she figured that rather than waiting on a lonely street for me, she would be better to tag along behind the couple that was just ahead of her, thinking that she would be home long before I left to meet her.

I was walking *up* Fort Street to meet her as she was following the couple *down* Watling Street, one short block over. She got to the house and found no one at home, the house in total darkness. She waited for quite a while, and then it became apparent (to her) that no one was going to be there—ever! She went around to the rear of the house (not having a key), got something to stand on and broke the kitchen window. She was able to get the window open, climbed in, cutting her hand in the process, put the lights on and the first thing that caught her eye was the little ceramic rolling pin that had been given to us.

I had left it on the kitchen table. Reading the little verse on it, she burst into tears, thinking that I had left her! The verse read:

"YOUR SLIPPERS ARE NOT BY THE FIRE, DEAR; AND HERE'S A BIT OF LATE NEWS. IF YOU CAN'T GET HOME BEFORE THIS, DEAR; YOU'LL FIND SOMEONE ELSE IN YOUR SHOES."

When I turned the corner to our street (Albion Crescent), the lights were on in our house. I unlocked the front door and Mary flew into my arms, tears rolling down her cheeks. It took quite a while to get her calmed down enough to eventually get the story from her. And guess who had to repair the broken window?

~ * ~

Two years later our second daughter, Shirley Christina Morrans, was born. She wasn't due until February 1959 but decided that she couldn't wait and so arrived at around five-o'clock in the morning of the 31st of December, 1958—seven weeks early. She was born at home, as this is what Mary and I decided (we could do that—our choice) after the carry-on we had at Motherwell Maternity Hospital during Audrey's birth. At that time, technology wasn't anywhere nearly as good as it is today, and apparently it was dangerous for a baby to be that premature.

It was fortunate Shirley chose the 31st of December which is New Year's Eve, called *Hogmanay* in Scotland. Hogmanay is about the most important holiday for us Scots. It was tradition for everyone to have a bottle of Scotch in the house at that time of year so as to be able to offer a "wee dram" to any "first footers" who may appear at the door to wish us a "Happy New Year." If it had been any other time of the year I wouldn't have had *any* whisky in the house as I didn't normally drink the stuff then!

The midwife was sent for shortly after midnight. She arrived, checked things and left again, saying that she would be back in two hours. She returned exactly as she promised. The midwife then worked with Mary while I did all the hard work (again!) of walking the floor downstairs! When Shirley finally arrived, she was blue—and that was not good. The midwife asked me if I had any whisky in the house. I said "yes," that I had a bottle. She ordered it and a basin, too. When I had brought her both, she laid the baby in the basin, opened the bottle of Scotch and poured *all* of it over the baby, massaged her with it. The midwife then told me I had to rush to the phone to call for an ambulance and oxygen immediately.

It was a one-minute run to the nearest phone kiosk (call box). There I found a button that could be pushed in case of an emergency. A male voice answered and asked me what I wanted. I told him I needed an ambulance and oxygen immediately for a premature birth as the baby was struggling for life. This idiot told me to go and find a policeman to verify my story. Well, I think I called that bloke everything under the sun and told him that if my daughter died I would hold him personally responsible!

The ambulance arrived at the house, took the baby away— not to Motherwell Maternity but to Bellshill Hospital, where she was put into an incubator. Mary was fine, as the afterbirth came away just before the ambulance arrived. Shirley came home after two weeks in the hospital and remained in excellent health.

(For many years I kidded Shirley about owing me a bottle of Scotch.) One day—maybe around 1995— she and her family were spending a vacation with us when Shirley came to me with a bottle of Ballantyne's. I asked her what that was for. She gave me a nice wee kiss and laughingly told me, "This is the bottle of Scotch I owe you, Dad."

Well, I gladly accepted it, not only because I didn't want to hurt her feelings but also because I had learned to appreciate a good whisky by then!

~ * ~

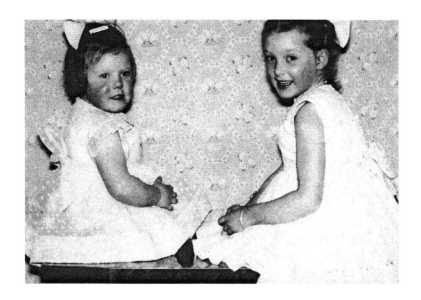

Our beautiful daughters, Shirley Christina Morrans, born December 31, 1958 and Audrey Susan Frazer Morrans, born August 13, 1956. I took the photo in 1962.

One of the jobs I had around this time, but not for long, was serving as a "tick man." This consisted of making my rounds throughout the local area to collect money from people who had purchased goods—clothing, furniture, etc.—and had taken it on "tick", which meant that they would pay a portion of the total every week. Interest, of course, was added on. (This sort of thing was probably the forerunner to the credit card.) It wasn't a bad job and there was a lot of walking involved so it kept me fit.

There was one house at which I used to call every Saturday morning. It included a little Pekingese dog that just hated me (and probably everyone else, too). It would get a hold on my trousers every week and I finally was fed up with it. (Although I love dogs, this one was too much!) I had to tell the woman to make sure that her dog was under control when I was coming. The situation was "all clear" for a few weeks. I found it quite nice to be able to go up to the door without this little nuisance trying to bite me. I always looked carefully for it possibly appearing before I opened the gate, only entering when I was sure that it wasn't around.

Well, one Saturday morning all seemed just fine, no sign of the dog. I opened the gate, entered, closed the gate and had just started walking towards the house when this little "Tasmanian Devil" came flying round the far corner, straight for me and barking its stupid head off! I quickly took a little side step and, just as this animal should have been trying to bite me, I got him under the belly with my foot and he ended up sailing right over the hedge that was along the pathway.

I was quite pleased with my shot until the woman of the house came around the same corner and saw me kick her little dog. I was expecting a blast from her.

Instead I heard, "Good for ye, Son, maybe that's what the wee rascal needs. By the way, I canna gie ye ony money this week. I'll try and make it up next week; is that okay, Son?"

I wondered what she would have said if she'd been able to give me her installment. I don't think that she would have been quite as nice.

~ * ~

Although Mary's mother and I were friendly, I still wanted us to get a place of our own. I had applied to the local council and one day got a letter saying that we could have a vacant two-bedroom flat in an area called "Forgewood." It wasn't what you might call a great area, but it wasn't too bad. Although we moved there, Mary kept asking me to see to her mother, as she really wasn't well. Well, this is where I really screwed up.

What I should have done before we moved was go to the local council and have Mary's mother's house put in my name. We could have stayed where we were to look after her mother, and the council would have had the flat to rent out to someone else. I wasn't "with it" as they say. I ended up taking Mary's mother with us to a two-bedroom apartment that was two floors up, and only half the house that we were leaving behind! Now *that* was really stupid! This was a real crossroad in my life later. Not the best of circumstances for Mary's mother; but maybe it was for the best in the end.

~ * ~

I wasn't earning very much as there was no overtime as a tick man, so I applied to the Glasgow Transportation Department for a job where there was always lots of overtime because of a shortage of workers. They were continually looking for people to be conductors on the busses or the trams and it seemed a never-ending process with them as there was a high turn-over of workers. Conductors were the people who collected the money for the fare and gave out the tickets for a certain distance traveled. 'Tram' was the British name for a street car. (They're all gone from Glasgow now, a long time ago.)

Because I had a little BSA Bantam, a 125-cc. motorcycle, and could find my own way into Glasgow (about 12 miles or less than 20 km.) in time to be able to "man" the first bus or tram out of the depot, around 4:30 a.m., I was hired.

After a period of training I was sent to work on the trams. It wasn't really a bad job. The hours were lousy, and sometimes we had to do what was called "split shifts." This was where a person would start work, have a three or four-hour break in the middle of the shift then resume to catch the rush-hour later. Working on the trams certainly couldn't be called hard work. Sometimes it was downright monotonous! Once it got quite dangerous for me.

I was on the late night run from Anderson Cross, doing what was called the "Saturday Wino Special." This was the tram that took all the drunks home from a notorious pub at closing time. Whoever was the conductor on the Saturday Wino Special was pitied far and wide. I was on the upper deck collecting the fares when a drunken man insisted that I have a drink from his bottle. When I told him that it was against the rules for me to have a drink while I was on duty, he got quite angry with me and insisted that I have a drink with him, almost forcing me to drink from his bottle. The drunk wouldn't take "no" for an answer, asking things like, "Ah'm no' good enough fur ye tae hae a drink wi'?"

He was very threatening and I knew I was in trouble; that is until a very large man came up to us, got hold of the bloke, just about lifting him off the floor and shouted at him, "Sit down, shut up, and if ye move a muscle I'll throw ye aff the tram; got it?"

Suddenly sobering up enough, the drunk sat down and I didn't hear a thing from him again. I thanked the man who came to my aid and his response was, "It's nae bother, Son; naethin' pees me aff mair than an idiotic drunkard."

Again, the money wasn't good at this job, but there were lots of overtime hours to make it better. I didn't mind working the hours as long as I was in there (in Glasgow). I certainly didn't want to go back into Glasgow for a couple of hours after I got home to Motherwell.

~ * ~

My tram was traveling east, "up the Gallowgate" one nice, sunny afternoon, heading towards Tollcross. I had collected all the fares and was standing at the back of the tram, facing straight up to Frank, who was at the driver's position, and not paying too much attention to anything.

The trams were completely reversible—when the end of the line was reached, the driver simply changed places and what was the rear of the tram then became the front, as there were controls at each end. Also, both the front and rear of the vehicle were wrapped around with glass so that it was possible to stand at the back and look ahead, straight out through the front window! (Also, there were "points" that were changed at the end of the line to get the tram onto the other line.)

Suddenly I saw the back end of a lorry parked across the tram lines quite a bit up the road from us. I saw that we were getting closer—and closer—and Frank didn't seem to want to stop. I figured he should be reducing speed and putting on the breaks any moment; but no, he didn't. It was a good thing he wasn't going very fast (we were a little early), for Frank drove straight *into the back of the lorry*! I think he was asleep at the controls, but wouldn't admit it (not even to me later in private). He insisted that he "just didn't see it." Maybe this was possible, but I don't think so.

As soon as the tram hit the lorry, the front of the street car caved in (as I said, it was mostly glass) and Frank was partly thrown into the main compartment with a good part of the front of the vehicle on top of him, plus the control box. After I saw that he wasn't seriously hurt, my main duty was to dash upstairs to the top deck and pull on the rope that disconnected the trolley arm from the power line. I did this and went back to Frank. The controls were what are classed as "dead man's controls" which meant that as soon as the hand is taken away, it automatically returns to "off," and this is what happened in the crash.

A transit inspector instructed me to go straight to the office in the tram depot to make out a report about the accident. I sat down with a pencil and paper the people in the office gave me, and wrote down exactly what I had seen. Well *that* didn't go down too well. I was told to state that I *didn't* see anything, as I was upstairs collecting fares from passengers at the time of the accident. I said that to make a report like that was telling a bunch of lies. That didn't matter to them; I was told to write that into my report, or I would be looking for another job; so I reluctantly amended my report.

I went to Frank's home to visit him a few times when he came out of the hospital, but never saw him back at work again. I don't know whether he was fired or whether he wasn't able to do his job because of his injury.

~ * ~

Another time while working for Glasgow Corporation, I was going down London Road heading towards Glasgow Cross with the first tram out in the morning. Two young conductresses from SMT (Scottish Motor Transit, a local bus company that travelled to various towns around the area) boarded. I think they were going to the bus terminus at Argyll Street and Renfield Street. Apparently there were two "busybodies" on board at the time (about 5:15 a.m.). They noticed that I didn't take any fares from the two young ladies, and they reported me for not doing so (a total cost of three new pence; or 10 cents).

A few days later I was told to report to the administration office and had to appear before the "king" himself—well, he must have been the king, for I'm sure he wanted me to bow before him. He asked me why I didn't collect their fares. I told him that it seemed the unwritten thing to do, as quite often I had been on one of their buses, in uniform, heading to where I had to "take over" on a tram somewhere other than the depot, and they hadn't collected a fare from me.

You would think with an explanation like that he would say, "Off with you and don't do it again." Oh no, nothing as sensible as that! "You are suspended for two days without pay," he said, with an air of authority that almost made me laugh. Oh, well, *another* two days pay down the drain. I really didn't mind. I had been working quite a lot of overtime and was actually in need of a rest. Stupidly, because of suspending me, they then had to put someone else onto overtime to cover for me, so it cost *them* money to punish *me*! (Daft folk!) All this for three new pence or ten cents!

~ * ~

Once during my tram tenure, I went outside before settling down for the night to put the padlocks and chains on my motorcycle's wheels to lock them. I opened the pannier bags (saddle-bags) and found the chains and padlocks gone. Further along the street lived "Blondie and her man." They had two kids whom we considered "right wee holy terrors." I figured that it was those two kids who were responsible, but couldn't go there and say so without any evidence. Instead I went to their door and asked Blondie if she would ask her kids if they happened to see any other boys take my things. (The chains were of no use to anyone as I had always put the padlocks through the chains and locked them, making them useless to anyone who didn't have the key.)

I've referred previously to people who "never worked nor wanted," people who knew how to work the system; well, Blondie and company were something like that. When I was taken into the house to ask the boys, I was stunned by the quality of the furnishings they had. Beautiful gilt mirrors on the walls, nice quality chesterfield and chairs, good area carpet, etc.—I couldn't believe it! As far as I knew, these people had probably never worked a day in their lives. They lived by their wits. When I got back to our house, I told Mary what I had just witnessed, saying, "Here's me, working as hard as Ah can tae get ahead, and Ah probably will no *ever* hae a hame as fancy as theirs." It made me rather disgusted. By the way, I never ever got my chains and padlocks back.

~ * ~

We were living in the two-bedroom flat when I bought my first little car—a 1932 8-horsepower Ford. It was a tiny car, could seat four and the spare wheel was bolted onto the outside of the back, for there was no boot (trunk.) The engine was so small that if I had wanted to I could have lifted it out of the bonnet (hood) myself. The bonnet opened by unhooking the side and folding it over, depending on which side of the engine was being worked on.

Nevertheless, it was a dandy little car and it took Mary, her mother, the baby (Audrey) and me on many a trip to Ayr or

Troon. These towns are at the coast and it was good for Mary's mother to get the fresh sea air. I didn't have much money and Mum would even help to pay for the petrol so that we could go somewhere to give her a wee change. I never asked for help this way, but when I went to the pump, she would hand me a few shillings, saying, "Here, Son, take this. Ah'll pay this time."

My little (really *old*) car didn't go down well with the men I worked alongside. (By this time I had quit the transit company and was working locally.) Often I would be asked, quite seriously, by one of them, "Who the hell do ye think *ye* are, wi' a *car*?" They didn't think it was right for a poor working man to try to "rise above his station." My reply was that I didn't go and stand in the bars every evening like they did, drinking beer and spending all their money by pouring it down the drain and then peeing it all out again. Nor did I bet on horses or race dogs.

No, it wasn't right that I should have a car that was about 25 years old! They were even unhappier the following year when I sold it and got myself an 18-year-old (1938) Hillman Minx, then the following year a (1939) Morris, then a year later a 1946 Austin! Then a year later I got a 1951 Triumph Renown, which was a nice big car, with leather seats and an all aluminum body. (Mary didn't like it because it "looked too much like a Rolls-Royce!") It really did, with its big chrome headlights, big chrome front and apron, and it was the same size too! Unfortunately it got the same lousy gas mileage, as well!!!

While I'm talking about cars, I might as well insert a humorous situation that arose with my Morris. The steering box was so badly worn that I could rotate that steering wheel a full half circle before the front wheels moved even a little bit. Deciding that I should purchase a replacement, I went to the local scrap yard (wreckers) and asked my buddy there if he happened to have a Morris car in his yard that I could take the steering column from.

"Aye, there is one. It's away at the bottom o' the yard towards the right. The car hasna any wheels on it so Ah don't know how ye are goin' tae get the steering column away from it."

I headed in the indicated direction and, sure enough, there was a Morris sitting on its axles (no wheels to lift it off the ground). I went up to the owner's hut to borrow a shovel, then started to dig a trench from the front of the car towards the rear and in line with where the steering box was. Then I had to get *under* it so I had to dig quite a deep ditch. (It's amazing what the lack of money makes a person do!!!) It was already evening when I had arrived and it took me quite a long time to get my trench dug and the steering column freed from the car. Good job it was the middle of summer with long days. I got the steering column onto my shoulder and headed for the hut in which the owner sat, to pay for it. There was no one there! I hollered, "Hello, anybody here?"

Even after I did this a few times, there was no response. I did meet up with the two guard dogs that were there to protect the premises during the night but they didn't bother me. In fact, I petted them for a wee while as I waited for someone to show up. Nobody did, so I ended up standing the steering column against the outside wall. I then put an empty oil drum tight against the wall, climbed onto it and lifted the steering column over the wall, dropping it down the other side. Lifting myself over the wall, I then headed for home. The next day I installed the "new" column onto the car.

A few days later I was in the vicinity of the scrap yard and called in to pay for my goods. When I saw the owner, I told him that I had called in to pay him the ten shillings that he had quoted. He was surprised to see me there and chided me on it. Well, after all I actually was a regular customer to his place and I felt I had to be honest. On parting from him I said, "By the way, Ah dinna think much o' yer guard dogs, all they did when they saw me was wag their backsides when I patted their heads." (Doberman's don't have tails!)

He just laughed and replied, "Aye, that's what they're supposed tae do—ye were already inside. Try that again—only this time try entering over the wall from the outside—ye'll get a different welcome!" (Yes, I guess he was right. I was already inside the premises—makes a difference!)

There was another quite funny situation with the Morris; well, at least in retrospect it was! The battery eventually went dead for good and for a time I couldn't afford a used one at the scrap yard. It so happened that the flat we had moved to was at the *bottom* of a short steep hill and the car was facing away from it. To combat the dead battery, I would remove the four spark plugs, put the gear shift lever in "reverse" and wind the car up this little hill *backwards* by using the *starting handle*. (Cars haven't had *them* for about a thousand years!) With the spark plugs removed it was very easy to do this, as when there was no compression, there was very little resistance and the car rolled backwards and upwards very easily.

When I got to the top of the hill, I replaced the spark plugs, would give the car a little push, hop in, put it into gear and off I would go to work. (It wasn't a lot of fun, as you can imagine, especially when it was pouring with rain; plus that, I had to get out of bed a little earlier.) I often wondered what the neighbours thought and if they had a good laugh while watching my antics every morning. There was a little hill at work where I parked, so there was no trouble in getting it started when going home, the *only* trouble was that I couldn't stop for anything on the way home unless it was on a hill, and it was just about *impossible* to start that car with the starting handle. (There was no parking allowed overnight at the top of the hill or anywhere close that I could have used.)

~ * ~

A few months after this I received word from my mother in Campbeltown that Bill Moorhead, my step-father, was very sick and had been taken to hospital. We decided to make the drive to see him, even considering the fact that it was the middle of winter. By today's standards it wasn't very far, only 150 miles (240 km.), but that was really considered a hefty journey back then.

I had changed cars by this time to the 1951 Triumph Renown. This was quite a big car; but my buddy, Ivor Riach, (deceased 1975) owned a little Hillman Husky, a much smaller vehicle which got a lot better gas mileage than mine did. He suggested that we swap for a few days, and I thought it was a great idea, especially as his car's rear seat folded flat. This would let me make a bed for our two girls, five and three years of age, and allow them to sleep most of the way. (There were no seat-belts in those days!)

It was quite late when I returned from switching cars, which meant that we would be leaving around midnight. While putting a few things in the car for the trip, I was surprised when a police car pulled in behind me and the constables asked what I was doing. I think they had the idea that I was in the act of stealing.

I told them what was happening and where we were going, and they were able to tell me to be very careful, as the last report they heard mentioned *snow* on the "Rest-and Be-Thankful." This is the nickname for the road that goes over a mountain called "The Cobbler" and is—well, used to be—a very treacherous one. (It has been altered to a great extent, making it much safer now.) This was the only route that would take us to the west coast of Scotland, so I had no choice.

I got the "rear seat bed" organized and both girls were tucked in nice and warm when we left home a few minutes before midnight. We travelled through Bellshill into Glasgow, along Argyll Street and onto Dumbarton Road, and from there to Tarbet with not too much bother. The first part of the journey was the "slow" bit, as most of it was through the built-up areas of Glasgow and Clydebank. Beyond Dumbarton it became more open countryside.

There had been occasional flurries, but at that point it became heavier; and as we were approaching Tarbet the visibility got much worse. Just beyond that village is the approach to the Rest and Be Thankful—so called because it was a severe climb—335 m. or 850 feet—and when you finally got to the top, you had to rest and be thankful you had made it there!

The snow was quite thick upon the ground on each side of the road, but the road itself was still fairly clear, though the roads of that time in Scotland were quite narrow (still are in many parts). We were heading upwards on an even narrower road, for the snow had already reduced the road's width. Adding to the problem, car headlights were absolutely *terrible* in those days—nowhere near as efficient as the "sealed beam" lamps that appeared *before* halogen bulbs came on the scene! The "old style" headlamps really weren't much better than two torches (flashlights).

The car was climbing and doing all right when I cried to Mary as I brought the car to a halt. "*Oh no! Look*!" I could see the headlights of a vehicle (I found out a little later it was a big lorry) coming down the hill to meet us. The unwritten law in Britain is that the vehicle that is *climbing* has to reverse to the nearest "lay-by" (a small section of road that is wider to allow one vehicle to pass while another is waiting).

To make matters worse, there were no such things as back-up lights on vehicles in those days. I don't know who the driver of that lorry was, but he *reversed up the hill*, to where the road was wide enough for me to pass. This was something he didn't have to do, and I loved him for it! *I* couldn't see to reverse *down* the hill, so how he managed to see where he was going *up* the hill, amazed me. This allowed me to continue to the top of the road that crested the mountain.

I had almost reached the crest when it stopped snowing and soon I was at the short, level bit before the road headed downhill again. I decided it was a good time to "rest and be thankful" while having a cup of tea. I then shut the engine off to gather myself together. I wasn't used to driving in snowy weather, and it was quite nerve-wracking—made worse for me because it was pitch dark. My insides were churning; it was two o'clock in the morning. We were away up at the top of a mountain pass and the only sound was the wind blowing.

Mary got the thermos flask out and produced some tea and a bite to eat. I felt a little better. We spent maybe a half hour there.

Okay—it was time to go. I switched off the interior light, turned on the ignition switch and pushed the "start" button. All I got from the starter was a groan. I tried again. Same thing. *The battery was dead!* I looked everywhere but there was no *starting* handle in that car! Away back then, all motor vehicles were equipped with starting handles (usually under the seat) that the driver could use, if needed, to start the engine. I thought, 'Now what can I do?'

As far as I was concerned, I was on top of the world and there was not a soul nearby to help. My wife and children were in the car and I had to get them to a safe place. The children were asleep. If I sat there waiting for someone to come along and help me I could have waited for ages. This was the middle of winter and not too many people travelled this road. I knew there would be a bus coming this way the following afternoon, but what would I do until then?

I didn't know what to do and I didn't want to tell Mary this. We hadn't clad ourselves in winter clothing like we learned to do later in Canada. Our clothing was light-weight and, besides, parkas didn't exist in those days. It didn't enter my mind that maybe we could all freeze to death, (although, at this later date, I now know that a good case of hypothermia was very possible). I *did* know we were in trouble.

The only thing I could think of was to put the gearshift in neutral, get the car rolling down the other side of the mountain, and when I had it moving, to hop in, depress the clutch, put it in gear and then release the clutch to start the engine.

There was a freezing cold wind blowing, but the clouds had passed so we could see the stars. They provided the only light I had as there was no moon; most everything was in darkness.

I got out of the car and put my shoulder against the door upright (on the right "British side" of the car), pushed, and gradually got the car moving. Then I suddenly thought, 'What if I slip when I'm pushing the car on this icy road, lose my grip and the car plunges down the mountain with my wife and children inside?'

I had to take the chance, for the car was then rolling pretty well and there was no way of stopping it. I managed to jump inside. I didn't know where the road was so I had to guess. I did what I was supposed to do and was so relieved when the engine started. But, as soon as it did, I had to jam on the brakes. I switched on the headlights and got out of the car. It was a good job that I *had* jammed on the brakes as soon as the engine started; for if the car had rolled just a few feet further, I would have been in serious trouble–much worse than we would have been had I remained where I was. The car would have been off the left side of the road, would probably have tumbled sideways down the hill and who knows where it would have stopped. I doubt if I'd have been here today to write this.

Have you ever thanked God and *really* meant it? I did! Even to this day, unless they read what I've written here, my family won't know just how much danger they were in. I didn't stop that car again until I got to my mother's place. And about the first thing I did the following day was buy Ivor a starting handle!

When we had eventually returned home and I related the story to Ivor, he was all apologetic. He told me that he had used the starting handle from his car for something and had left it lying in a corner in his kitchen!

So now he had two, because the handle I bought for his car would fit *only* his car, no other; so it was no good to me!

~ * ~

Mary's mother, Shirley, Audrey and I on holiday at Dunoon pier on Kilbrannon Sound of the Firth of Clyde, 1960. I guess Mary is not pictured because she took the photo.

The flat we had in Forgewood was not good for Mary's mother. She could hardly get a breath. Nor could she manage the stairs. Our ground-floor neighbour, George Bell, would help me whenever Mum had to go out somewhere, such as to visit the doctor. George and I would make a "chair" for her by clasping our hands together under her, so that she could sit on them. Then she put her arms around our necks and we carried her down two flights of stairs in her sitting position. If George was at work, a soccer game or elsewhere, then Mum had to stay home.

I wrote a letter to the town council explaining just how she was a prisoner in her own home and asked if it would be possible for them to give us a house with a front and rear garden so that she would be able to get outside to get some fresh air. No reply. I let a short time go by and wrote to them again. Still no reply. I wrote again, asking if I could talk to someone about her situation, and I got a reply that said I could meet with the person in charge of housing.

When I was finally talking with the housing boss, explaining about the difficulty of getting Mum out to get some fresh air (she had severe asthma and a bad heart), he said something about people getting a better house because of the parent being unfit, and after they were issued with the house, they would put the parent out to a nursing home. I assured him he needn't fear anything like that happening to Mary's mother, as, if it meant there was a possibility of getting a house; I would let them put the house in *her name*. He was quite surprised with this offer and asked me if I meant it. I said that I most certainly did.

A few weeks later we received a letter telling us that we had been allocated a newly built townhouse at 52 Coldstream Crescent, Wishaw, and it was in *my* name. This was a dandy house. It had three bedrooms upstairs, a nice big kitchen and a 21-foot long living/dining room combination going from the front to the back, with a large picture window at each end. It was really modern and had *electric under floor heating*. (So I can safely say that we had central heating while we were still in Scotland in 1963.)

There was a nice unfenced front yard and a chain-link-fenced rear yard. And it bordered on a forest at the back. The rent for this property was two pounds sterling (six dollars!) a week, about eight and a half British pounds or 25 dollars a month (10% of my income). This was great for Mary's mother. Sometimes we would have her bed down in the living-room and sometimes it would be upstairs, depending on how she felt. She could sit at the front door or the back door, to get some fresh air. (I managed to carry her down the one flight of stairs on my back.)

~ * ~

Here I am in my Glengarry bonnet posing with baby Shirley
after a Cameronian Military Band performance, 1959.

During all this time (from 1959 to 1964) I was in the Cameronian Military Band of the Territorial Army (known elsewhere as the Militia—part-time soldiers or "weekend warriors"). This was a very good band with quite a few first-class players (and I wasn't one of them) among the 48 bandsmen. I got on to being on "first cornet" but was nowhere near good enough for solo cornet. No, that position belonged to John Davy who was one of the best trumpet players I've had the privilege of knowing. The first trombone was excellent, too. This was Ivor Riach, my buddy who'd lent me his car. Ivar could almost make his trombone speak. He was asked to go professional four times that I know of by a big English dance orchestra. The solo euphonium player, Peter Lindsay, was another first-class player. "Blackie" (Blackwood was his real name) was tops on clarinet and Stewart on French horn.

Then there was our illustrious bandmaster, Mr. Jimmy North. This man was so far away in front of any bandmaster I've ever known; that it wasn't even funny. He was musically educated at Kneller Hall in England, the place where all the topnotch bandmasters go through their stuff. This man is the only person that I've met in my life who had "perfect pitch"—the ability to name what note was struck on a piano while being in another room and not being able to see which key it was. He could sit at the radio and listen to a "top of the hit parade" pop tune number and write the notation down while the tune was being played! This enabled us to be able to march along the street playing pop music. We "brass band diehards," so used to marching and playing stuff like "Colonel Bogey," "Stars and Stripes Forever" and the like, didn't really relish this kind of music, but the folk at the side of the street really enjoyed tunes like "Hard Day's Night" by the Beatles.

It was probably illegal for Jimmy to do this but he didn't seem to worry about it; he wanted to please the crowd and that's what he did. He would have parts written out for all the instruments in no time. Utterly fantastic! Unfortunately, like most extremely gifted people, he had a poor sense of humour. In recognition for his work in the music field, he was awarded the MBE (Member of the British Empire) medal and had to go to Buckingham Palace to receive the award from the Queen.

These people I've just mentioned were all ex-regular fulltime Army bandsmen. They all had joined up as "band boys" and had been trained since 15 years of age, and I mean *thoroughly* trained in playing their individual instruments, probably six hours solid every day, for 12 years. When you do that you get to be either an excellent player or they put you out of the band.

It was while I played in the Cameronian band that Ivor got me into the Lanark Loch Dance Band. This was a really good 17-piece orchestra that played in its own hall on the outskirts of Lanark. We played every Saturday and Sunday night from eight o'clock until midnight. The wee bit of extra money came in handy.

~ * ~

Cameronian Military Band (Territorial Army), Hamilton, Scotland, 1964. I'm fifth from the right in the middle row, holding a trumpet. Territorial Army photo.

The Territorial Army Band had to go to "camp" for two weeks every year. This was a fun two weeks with lots of rehearsals, "Beating Retreat" in certain towns (an evening ceremony at sunset, and nice to watch) and concerts. Unfortunately, it was during the last camp I went to that my mother lost her husband. Bill Moorhead had died. When she found out, Mary had tried to get in touch with the Territorial Army so that I could get home but, due to incompetent people, the message was never delivered to me. I caused quite a stink at Army Headquarters, telling them what I thought of their administration, which couldn't get a message to me to tell me that my stepfather had died. That was me finished with that lot!

Mary had been at me for some time to give up the band, as nearly every weekend was taken up with playing somewhere. I didn't seem to have any free time for the family. The band would start weekends by playing at just before kick-off and at half-time at a football (soccer) game in Glasgow on Saturday afternoon, then we'd play at the dance hall that evening. I had to be at the band hall by 10 a.m. Sunday morning for rehearsal until four p.m., and then back to the dance hall that evening, too. After the Territorial Army administration had let me down when my stepfather died, I didn't feel like having anything more to do with them; so it wasn't too hard to finally agree with Mary's request.

For a time after that, most of my music-making consisted of whistling around the house or setting one or the other of my daughters on my knee and singing my version of an old Scottish tune that originated with a sweets peddler in the 1800s and had recently made a comeback:

"*Ally, bally, Ally, bally, bee, Sittin' oan* yer daddy's knee* (the original said "mammy's"), *Greetin'* for a wee bawbee*, Tae* buy mair* Coulter's candy.*"[7]

*Scottish words: "oan" (on); "greetin'" (crying); "bawbee" (half-penny); "tae" (to); "mair" (more).

~ * ~

---

7   First verse of "Ally Bally Bee" or "Coulter's Candy", by Robert Coltart (died 1890).

Shirley (3) and Audrey (6) at Dunoon Pier, Summer of 1962.

Shortly after this I was talking to Gordon, a neighbour two doors away from me who worked for Colville's, a large steel company. I think he was in charge of a department at the head office of another branch. He asked me if I knew anyone who would like a job as a clerk in the new Ravenscraig Steel Works that had opened just a few years before, was expanding, and was now getting ready to produce coils of sheet metal. I told him that I didn't know of anyone. That was when he happened to mention that the wage for a clerk there was 11 pounds, 10 shillings a week. Hey, that was two pounds a week *more* than I was presently earning, so I told him that I did know someone— me. He said that he would arrange for me to have a talk with Mr. J. Dixon, Assistant Comptroller, and to leave it to him and he would organize an appointment for me.

I went to see Mr. Dixon at his office after Gordon had arranged a meeting. Mr. Dixon then told me that one of the biggest problems in the job was reading what someone else had written. It seemed that poor penmanship was quite a problem, so in order for me to apply for the job, I first had to write the numbers from 1 to 10 and also spell each one out in longhand and then print the words in capital letters. I did this and he took the paper and examined it. "Very good" he said, "at least we'll have *one* person whose writing is legible. When can you start?"

That was it? That was the test? I couldn't believe it. I started the following Monday morning.

I didn't know it at the time, but this Mr. Dixon (Jimmy) lived just a few doors from both Gordon and me. Gordon was on my left and Jimmy was on my right on going out the front door. It did turn out quite well for both him and me, because there were times when he would be called to the office during the evening to sort out some problem at the rolling mill, and, instead of getting the bus, he would ask me if I would drive him there and give him some help in locating some things.

I wasn't able to give him much help at first for it was all new to me, but I gradually got to know what was going on and soon was able to help him without him having to explain what I had to do. Sometimes I was able to make a suggestion. I got overtime for this, plus I learned quite a lot about how the sheet steel rolling mill operated. It wasn't long before it paid off. There were two shift schedulers, Angus and Bill. I was assistant to Bill and we were on two shifts. The situation expanded quite rapidly and one day I was told to report to Jimmy's office. Once there he told me that he was very pleased with the way I had helped when I didn't have to and he wanted to know if I would be interested in becoming a scheduler for a third shift. There would be a good increase in wages. "Sure," was my eager reply. Hey, "Yours Truly" was getting up the ladder a wee bit. I then had my own assistant and a clerk.

The wages were the best I had ever earned, which meant that I could now afford a *better* car. A few more people now had cars, so it wasn't as big a "crime" for Ian to have a car as it used to be! I got myself a 1961 Renault Dauphin. It was now early 1964 which meant that I now had a *three-year-old* car. But, just think, I wouldn't have been in this position if I had arranged to stay in the house that Mary's mother had instead of going to the flat and taking her with me. I did say it was a *major* crossroad.

~ * ~

By this time I had also bought Mary a wringer/washer and also a little refrigerator (they were *all* little, about the same size as a modern clothes dryer). Some of my workmates were kidding me on, because now I had *bought a fridge*! Very few people had one of *them*. I was now going through the same thing with a fridge as I did with my cars!

The washer didn't cause too much of a stir as more people had them than had fridges! I was likely a bit of a rebel against the norm in a way. I didn't go to the pubs and hang around with blokes that did that sort of thing; also, I didn't go to the "bookies" to bet on horses or the dogs. All the chaps I knew would go and stand in the pub and drink *warm* beer. I didn't really like warm beer so I would buy a couple of MacEwan's "screw-tops" (16 oz. bottles, a *true* pint), take them home and put them in the fridge so that there would be some in the house when anyone came to visit. I think this was unheard off in the early 1960's in Scotland!

It did pleasantly surprise any visitors we had, when during the course of the evening we would be sitting having a chat and I would say, "How would you like a glass of cold beer?" It sure did throw them for a loop, for I didn't know of anyone else who would buy beer away back then to take home. By this time I had all the "mod-cons" except a telephone. Boy, if I had managed to afford a phone, I would *never* have lived *that* down. Telephones were only for the gentry!

* ~ * ~ *

The family at Mary's sister's first wedding, 1962.
Our daughters were flower girls.

CHAPTER SEVEN
The Canadian Connection

Mary's cousin, May, was back home on a visit to Scotland with her husband, Bill McIlroy, a Canadian. It was Bill's first time in Scotland after marrying May several years earlier. Our get-together was in the spring of 1964 during one of the many parties that were being held in their honour at various homes, this particular one at May's brother William's and his wife Annie's place.

Though Mary and May were cousins, they were not close. I had not met May previously as she had immigrated to Canada quite a few years before I met Mary. We had, however, visited May's mother and father (Mary's Uncle Tom and Aunt Sarah) an odd time. I know that I had previously mentioned to Tom and Sarah that every once in awhile I had thought of eventually immigrating to Australia. I don't know why really—perhaps I just had a yen for a real change and thought the big wide world out there might be passing me by.

Over the many years that have elapsed between then and now, I have often wondered if our invitation was *engineered* in some way, maybe in such a manner as to give May and Bill a chance to talk to us about Canada before they headed back home again. Maybe there was a possibility May's mother had told her that Maisy (Mary's nickname) and her husband (that's me!) were thinking of going to Australia, and the suggestion then, perhaps, would be, "Let's get them to the next party and we'll talk to them about coming out to be with us in Canada."

225

I preferred to think that there wasn't any collusion, but out of all the parties that had been held, this was the only one to which we from the "other side" of the family had been invited. It happened to be the last one before the McIlroys headed back across the Atlantic a few days later. (I had often wondered, though, over the years, just how much different, for the good *or* the bad, my family's life might have been if we hadn't been invited to *that* particular party (another crossroad). I decided a little later that it was best not to interfere in what people wanted to do, even if they asked for advice. If they wanted to go to Timbuktu, then that should be up to them; others shouldn't try to talk them out of it. It's their life; let them lead it in whatever manner they wish. Then they won't be able to blame anyone for their fate, or look back and point a finger, saying that they were given bad advice—as I have sometimes done in the recent past!)

~ * ~

Bill and I were sitting together talking about how he was enjoying his holiday and how much different it was in Scotland compared to Canada. Then May took an envelope out of her handbag.

"Here we are," she said in a little louder voice to get above the conversation and laughter that was going on. (She already had lost most of the Scottish vernacular.) Turning in her seat to look at Mary and me, she was sitting on the other side of Bill. "Here, I brought these along for you to look at. Mother was saying that you two are thinking of going abroad, so maybe you should take a look at these photographs. It'll give you an idea what it's like where we live, in Pickering, on the outskirts of Toronto."

I took hold of the photographs that Bill passed to me and casually glanced through them, and as I finished with one I passed it to Mary, until we had looked at all of them.

"Aye, it seems very nice there; looks a wee bit like ye folks are having a grand time."

The first one showed Bill in the backyard, relaxing on a lawn chair with a bottle of beer. Next May was standing in the shallow end of a swimming pool and waving at the camera. The next one showed the two of them with May's brother, Jim, and his wife (they had followed May a year after she had left), sitting around a picnic table having a good time and holding up their drinks in a sort of salute to, most likely, their mother and father in Scotland.

Most of the remainder were much the same, all showing the beautiful sunshine, the flowered bushes close to the pool, the carefree lifestyle, then a view of their nice bungalow, and, of course, there had to be a photo of Bill's one-year-old big American car—a "'63 V8 Pontiac," he said it was.

"If you want to improve your life and your family's too, then Canada's where you should be, Ian. There's a lot more opportunity there than here."

My broad Scots west coast dialect came in again. "Well, aye, Ah can see it might be good there, but Ah was actually considering immigrating to Australia. Ye see; all of us, Mary, the two lassies and masel' (myself) can go there for only ten pounds (thirty dollars), and if we like it and stay there for two years, then the Australian government cancels the cost of the remainder of the fare. We won't have tae pay ony mair money at all. Dae ye no think that's very good?"

It seemed as though May was ready for that. "Oh, that bit's okay, but why do you think it's so cheap to go there? It's because it doesn't have the same standard of living that Canada has. You don't want to go *backwards*, do you? There's plenty of work where we are, and you will be much further ahead than going to Australia. Besides, what if you don't like it there? You'll be on the other side of the world and it'll take you more than the two years to save enough for the fare home, and that's only if you can find a job."

Bill nodded in agreement with that and then he picked up one of the photographs, the one showing his car. "Another thing to consider is that there aren't cars like this for the ordinary Joe in Australia. All they've got are old cars from England, and that is typical of the situation. Most of the stuff that's there is second hand from other countries. Listen to this now–one and a half hours drive from our house takes us into the United States, or the same time down to Niagara Falls. Imagine driving *there* just for a day's outing!"

Well, I was certainly not prepared to talk about my plans for the future. To start with, I really *didn't* have any plans to talk about! What I had was called "itchy feet," I suppose. All I really had in my head was a little thought that it would be good for the family to get away to a new country, and Australia was what I first thought of. I knew that there were kangaroos and some other very unusual animals there, that there were palm trees and miles of silvery sandy beaches, that their summers were our winters and Christmas was in hot weather, but not much more than that. Boomerangs too; yes, that's where boomerangs came from. It seemed that whenever I had heard of someone going overseas, it had always been to Australia, so that was where I thought we should head.

~ * ~

To be truthful, there really wasn't anything wrong with the way things were going for us in Scotland. The following sums up the general way of life for me and my family during this period: We were in a brand new, council-owned, townhouse in Wishaw. I had a good job at Ravenscraig Steel Works in Motherwell as a scheduler for the sheet steel rolling mill. I worked in a large, all glass, air-conditioned office building at the entrance to the steel works, and had an assistant and a clerk under my command. Looking back over all of those years, I can definitely say that it was the best job I ever had. I drove to work in my three-year-old car wearing a suit and tie. At the office I had an outside phone, an "inside-only" phone to all the departments, and was responsible for making up the "rolling schedule" for the following shift down at the mill. I was on what was called "continental shifts" which meant that my two days off continually changed. I was earning 33% more money than the tradesmen in the mill, so life wasn't at all bad for our family.

My salary was paid directly to the bank and I was asked by my bank manager to use a cheque book to do my business. This was practically unheard of in those days by "us" working class chaps. When I went home and told my wife about this she burst out laughing, saying, "Huh, imagine *us* wi' a cheque book?!"

After getting my cheque book, when I paid my bills and wrote out a cheque, well, you should have noted the dignity that was afforded me. I was treated like I was really somebody! I didn't know of anyone else that wrote cheques in 1964.

Though my wife didn't *have* to work, she had a part time job at a grocery shop in Wishaw. There wasn't a lot of money left at the end of the month, but there was enough for the family to live in a reasonably decent fashion. One thing is for sure, I was better off than I'd ever been in my life. There were no credit cards in those days, although we did have an account with an outfit in Glasgow to get an occasional thing on credit. But mostly everything that was needed had to be saved up for. Our only real debt was a bank loan for the car.

~ * ~

When driving home from the party later that night, I asked Mary what she thought of the conversation with May and Bill, noting that she didn't join in and that she should have; for it seemed that they had collared me every chance they got that evening. Not that I really minded, for it was interesting to find out about Canada, and they made it sound so nice. They also made sure that Mary and I had a piece of paper with their address on it before they left that night, telling us that if we ever decided to make the move to Canada that we would be welcome to stay with them for a little while until we got settled.

"Oh, Ah didny ken whit tae say tae them, Ian. Ah ken fine May's my cousin, and we were friendly enough wi' each other, but not really that close when she lived here. Ye know whit ah mean? Dae ye no' think that they could hae a point in sayin' that Australia's awfy far away? Maybe we'll never manage tae get back hame for a holiday if we were tae go there."

"Well. Ah'm no too concerned aboot that, m'dear, for the way ah see it, oor hame'll be oot there, no' here."

"But, whit aboot ma mother? Ah'll never see her unless we can take her with us. Ah canna go anywhere an' leave her tae hersel'; she'll never manage. Besides, did May an' Bill no' say that the Canadian government will pay oor way oot there, an' we can pay it back at 'so much' a month once you find work?"

Since we had reached our house the conversation slowed at this point and was to be continued inside.

"Dae ye want a wee cup o' tea, Mary? Then we can talk a wee bit more if ye want tae." Mary's mother was upstairs in bed and everyone was asleep as it was rather late; but yes, a wee cup of tea would be just fine.

The most logical reason for this is simply because the Brits discuss all-important things over a cup of tea! Our tea was soon made, and we enjoyed it along with a few biscuits.

"Ah don't think that we'll hae tae worry too much aboot the lassies, Mary. They're still gey (quite) young an' it'll no' concern them too much. They'll probably think it a smashin' adventure!"

"Aye, ah ken fine. Dae ye think ye can drive in tae Glesga (Glasgow) on one o' your days off, go tae the emigration place and see what ye can find out?"

"What are ye thinking o', m'dear. Dae ye want me tae go tae the Canadian Embassy or the Australian Embassy or tae baith (both)? I think the Australian wid be oor best bet."

"Ah'm jist thinking o' whit the McIlroys said aboot there being plenty of work where they are. Dae ye no' think that's whit we should be thinking aboot? At least we'll know somebody there, and they *did* say that we could stay wi' them for a wee while. That wid be a big help. If we go to Australia we won't know a soul. I think, tae, if ye get a job earnin' half decent money, ye can be reasonably happy no matter where ye are."

"Well, that's no' right, Mary, for Ah'm earning' a pretty good wage here, so why do Ah want tae move? That bit o' your conversation dizna make much sense! Besides, it's no' us that'll get maist o' the benefit, it's the waens (wee ones). We hae to think o' them. But there must be work in Australia, otherwise they widna be asking for immigrants, wid they? Whit if they got a lot o' folk tae go there an' there wiz nae jobs for them? It would be disastrous! They widna be able tae advertise in the paper if the word got oot that the folk were idle when they got there. They would be ridiculed for advertising wrongly, wouldn't they? Besides, it's no' as though we were thinkin' o' going tomorrow."

"Well, we canna talk any mair until we actually decide whit we're gaun' tae dae. Ah still think that ye should enquire at the emigration places an' we'll make up oor minds later. It'll surely be a lang time doon the road afore we're ready tae go anywhere."

"Anyway, we canna think aboot anything until we know that we can take your mother wi' us. If they won't allow that then we just canna go anywhere, either Canada *or* Australia, for we're certainly no' gaun tae lea' her wi' anybody else, that's for sure!"

I don't know why, but I didn't get around to going to either of the immigration places right away. But our reason for *not* emigrating soon changed quite rapidly. Mary's mother's health deteriorated and she died a few months after that party. I've often wondered if she had overheard us talking about emigrating and was under the impression that she was holding us back and lost the will to live. I really don't think so, as it was something that we didn't talk about much; but she was the kind of person who wouldn't want to be in the way. Still, one never knows.

As Mum's health gradually got worse, we decided to get her downstairs to the living room to be with us, so that she could have some company instead of being stuck in her bed upstairs. Plus that, she could see some television. She wasn't downstairs very long, though. I had to carry her down on my back, and I've often thought that maybe the extra effort of holding on to me may have been partly the reason of her demise. Her heart just gave out.

However, no more was said about emigrating, even quite a while after Mary's mother had passed away. Then one Saturday evening we went to visit Aunt Sarah and Uncle Tom, May's mom and dad. During the course of the evening, Aunt Sarah said, "I dinna know why ye two are still here; you should be away in Canada beside May and Jim."

Well, that started us thinking and talking again of going abroad. Mary's mother wasn't a concern to us anymore. We would rather have had her with us, of course, for she was a dear, sweet person, but the difference now was whether it was Australia or Canada. We no longer had anyone to worry about other than *just us*.

~ * ~

The question of Australia versus Canada was very difficult to settle. I eventually got information from both immigration places and then came the ironing out to see which would be the best for us. The Australian brochure told us that when we arrived there, the men and boys would have to go to a men's hostel and the women and girls to a women's hostel for a short time. We didn't like that bit too much (but it was the absolute truth) and the Canadian brochure didn't mention anything like that.

"Wh't d'ye think, Mary; where should we go?" Neither of us knew for sure. We had been writing to May in Canada (at least

I had, not Mary, who should have) and she always responded very quickly. One letter I got much later told me that I had to hurry up to get to Canada as her brother Jim had secured a job for me at the steelworks he was employed at. I think that was what initially settled the issue, sort of steered us in that direction, and the wheels were then set in motion.

* ~ * ~ *

The family on holiday at Whitly Bay, England, 1964. Our car was a 1963 Renault.

CHAPTER EIGHT

*Big Decision Coming Up:
Canada or Australia?*

The man at Canadian Immigration in Glasgow seemed very pleasant. "I see you are considering going to Toronto, Mr. Morrans. I think you should make a visit to the Government of Ontario office in West Nile Street. The official there will be able to give you a lot more information about the area than we can." I already knew that Toronto was in the province of Ontario, so his suggestion made sense. I gave them a call to make an appointment and made arrangements to visit there on my next day off. Meanwhile, I had more literature to read, plus I had picked up quite a thick book, all about Canada, which would give me lots to read about. It is one book that I wish I had kept close to me. I loaned it to someone and never got it back.

~ * ~

The Government of Ontario official welcomed me into his office and offered me a seat. During the course of our conversation, he told me about all the good things regarding Ontario and Toronto, and that it was, by far, the richest and most populace province. In addition, he told me, "as much as Canada wants you and your wife, it is really the children we want most, as Canada needs children to build the future." (I cringed when remembering this later as we encountered just the opposite view.)

I wondered (much later), if all provincial representatives took a training course on how to stretch the truth, even to the extent of downright *lying*, to encourage people to settle in their respective provinces, irrespective of whom it may influence wrongly, even hurt or damage newcomers, in their quest to gain population. It might have been interesting to visit officials of other provinces, if they existed in Glasgow, just to see exactly how each portrayed their individual areas.

This is the gist of the conversation between the Ontario official and me. Not verbatim, but pretty close. [I've added my retrospective comments.]

"See that smog out there? Well you won't see any of *that* in Canada." [Toronto is terrible *as I write this!*]

"Really?!"

"How far is it from here to Edinburgh, Ian?"

"Oh, I think it is around 45 miles."

"What would you say if I told you *that* is the distance from one end of Toronto to the other?"

"Really—that big?" [True, for Greater Toronto is even bigger today.]

"Yes, and you know something else? You can drive all the way through it and not see one slum."

"Gee, really? ... That's great! Wow! ... Now it says in this book I have here," *(I had the book with me that told all about Canada and held it up to show him)* "that at least 68 percent of the population either own, or are in the process of owning their own home. What's the possibility of me getting a house of my own?"

"Oh Ian, it's so simple. All you'll need is two hundred dollars to put down and you can walk into a *beautiful* house. Far better than anything you'll *ever* own here in Scotland!"

[Here I have to mention that in the first bit he was quite correct. You *can* drive from one end of Toronto to the other end and not see one slum. All you have to do is remain on the #401 Highway and you definitely *won't* see a slum, you *won't* see much else either, mind you! But come off that highway, go drive through downtown Toronto and you'll see lots, not just slums but absolute filth. Bad enough to make Steele Street in Glasgow look good. Lorne Street in Campbeltown was an "elite" district compared to what I later saw in the eight and a half years that I lived in the Toronto area.]

In the matter of the houses, well, I'm afraid that subject will have to wait until later in the story. He did tell me lots more, but I'm afraid that time has robbed me of most of it, but the general picture was that anybody who even *half* tried to improve was bound to do so. He asked me what kind of work I was going to do and I told him, also mentioning that I had a wife and two little girls to see to and that I would even go on the "rubbish" lorry if there was nothing else. That was when he said that it was almost impossible to get one of *those* jobs [another truth!] as they were all taken, but that shouldn't concern me as there were so many jobs vacant that it would be a long time before I would have to worry about that.

I was quite pleased with things. Everything seemed to be working out all right for us. I went home to Mary, just about floating on a cloud.

"Mary darlin', it's all settled. We're going to Canada and Ah'm going to buy ye a nice house shortly after we get there." It was then I explained the conversation I had with the Ontario government official in Glasgow.

"So, does that mean we definitely *are* going to Canada, Ian?"

"Ah think so, Mary. The way Ah see it is that if we go to Australia, we'll have tae live in the hostel they mentioned— and if we go to Canada, Ah'll be able to buy ye a nice house in just a few weeks. I dinna think there's any comparison there, do you?" She agreed and a decision was then made. We were going to head for *Canada*. (Hey, watch out for that *new house* a-comin'!!!)

~ * ~

There was a problem. We had to have medical examinations and an x-ray showed that Mary had a spot on one lung. We had already *told* them that this existed, that it had been caused by pleurisy many years before and to expect it to show up on the x-ray. They hadn't listened when we first told them about it. I've wondered for years why it is that people don't listen when you're trying to tell them something. Perhaps they think that what *they* have to say to *you* is more important than what *you* have to say to *them*. It took quite a while and a lot of running around to get it all sorted out, getting the original x-ray from Mary's own doctor, but we *did* eventually get them to accept this.

~ * ~

"So, Mr. Morrans, that is all the paperwork completed for you and your family. You are now able to emigrate to Canada anytime after you make arrangements for travel, when you decide which manner you want to go, whether by air or by sea. The good thing about going by sea is that you get 25 cubic feet of space for each of you in the ship's hold, as the brochure told you, so you would have 100 cubic feet for the four of you. It depends on just how much of your household goods you wish to take with you."

With the news that May's brother Jim had secured a job for me, I made enquiries about how soon we could make the journey. An extra ship had been laid on because of an overflow and I was told that her sailing date was November 2, 1965. This was good, I thought, for it had originally been planned that we were going to make the trip during April of 1966. We figured that if we were going to go we might as well go right away, since there was a job waiting for me.

~ * ~

I had recently bought a brand-new piano. Not that I could play well, just a little by ear, but as we were sending Audrey to lessons, thought we might as well get a decent one for her to practice on. Moreover we had just bought a new bedroom suite for each of the girls. Audrey's was teak and Shirley's was walnut. (Each had a headboard and footboard for the bed, a dressing table, a chest of drawers and a wardrobe.) This was only a few short months before we first thought of going abroad. If we had only known that we would soon be selling them we could have saved ourselves a whole lot of money—which we badly needed for our arrival in Canada.

This was all really nice furniture that we had to sell before we made the move, and we didn't have the pleasure of owning anything *close* to the same quality later in Canada. There's no way we could have afforded it. The two doors of the walnut double (his and hers) wardrobe that was part of *our* bedroom suite (purchased when we moved into the single-end in Dalziel Street) had a beautiful likeness to a butterfly's wings when both doors were closed. This was because the veneers on both were from the same piece of wood, except that the opposite side was reversed to create the pattern. Furniture of that quality would cost a fortune in Canada.

Bear in mind, too, that we had to sell *almost everything*. This goes from something as small as a toaster (anything electrical couldn't be used in North America), up through all of our household goods, everything that we had sacrificed for, and we got just about nothing for them. It was no wonder that Mary cried her heart out when we had to sell them. I can tell you right now that if I had to do it all over, I would say "no," for I wouldn't have put her through all that again, nor me either, come to think of it. Everything we had worked for was about gone. Eleven years of work and just about all we had left were bedclothes, personal clothes, dishes, pots and pans, some ornaments, and, of course, our two girls. If we hadn't had them, then it definitely *wouldn't* have been worthwhile.

It certainly would have made things a lot easier for us if we'd had *nothing* when we lived there, but we later found ourselves leaving a very satisfactory situation to find ourselves transplanted to where it took us *five years* to get back up to roughly the level we were at *before* we left Scotland. Of course, while this entire trauma (and believe me, it *was* a trauma) was going on in our lives, we are telling ourselves, "It is okay. We're going to a land that's *flowing with milk and honey*," so to speak, according to the bloke at the Government of Ontario office. Not *exactly* his words, but pretty close.

We had never *ever* bought any used furniture. In fact we had never purchased *anything* in our married life that had *ever* belonged to anyone else, except cars. (And the only used thing that *I* had ever bought in my life was the camera and tripod already mentioned.)

We soon found out that when anyone is selling furniture you have to be prepared to almost give it away, especially if it were known that you were going overseas. No one wants to pay anywhere *near* fair market value. If they learned that we were emigrating, they thought they had us over a barrel. Also, it is just about impossible to sell a piano. We had to sell at bargain prices and all I got for that beautiful piano was 20 pounds (60 dollars), which was a crying shame! All in all, what we got for all of our furniture almost left us with too little money to make the move! However, we were past the point of no return. The moving arrangements had been made, and as the ending for the last chapter said—"the wheels were in motion." We were fast approaching "The Big Step". *All I hoped was that it was going to be worth it.*

* ~ * ~ *

Farewell visit to see "Campbeltown Gran," which is what Audrey and Shirley called my mother. Campbeltown harbour, 1965.

CHAPTER NINE

The Big Step (And It's Bigger Than You Think, Boy!)

I had to gather "tea chests," wooden boxes that were originally used to ship tea from "wherever" to Britain prior to it being packaged for retail. They were eventually used by people like me to pack things into prior to moving. They were made of thin plywood stapled onto a one-inch square wooden frame that made up the "skeleton" of the box and there was a preformed metal reinforcement at each corner. The chest was about two feet by two feet by two feet, which made eight cubic feet.

As we were allowed to have 100 cubic feet of cargo space, that meant that we could have 12 of those chests, unless we used different-sized boxes along with the chests.

Another thing we had purchased almost immediately prior to making the big decision was a long carpet for the living/dining room. This carpet was top quality Axminster and I was offered "buttons" for it by Gordon, the chap who got me the job at Ravenscraig. His house was identical to mine, and as it was a carpet designed to have a "surround," then it would have fitted exactly into his home. The carpet was 20 feet long (about six and a half meters) and this bloke wanted to give me 20 pounds for it (60 dollars!).

I told him to "get lost" and he answered, "Ye have to sell it tae me, Ian, and it's really nae good for *anyone else* unless they cut it. Let me know when ye are ready tae sell and Ah'll take it off yer hands."

I replied, "Well, it is good for *us!* If Ah dinna get a buyer for it at my price, Ah'll take it *wi'* us!" And I did! I managed to get hold of a large, long crate, rolled the carpet up as tightly as possible, folded it over (which wasn't easy) and put the carpet into the crate along with some other things to help fill it. So Gordon *didn't* get the deal he was so positive that was coming his way. What it did though, was cut down on the amount of tea chests I could take with me.

~ * ~

We were going to Canada on the *Empress of England*, 25,500 gross tons. I had to make a trip into Glasgow to the Canadian Pacific office to arrange for the journey "over the pond." The people there were very helpful in advising me in regard to what was required for the trip. When I was discussing the position of where to pick a cabin for most stability, a lady that worked there brought a great big plan of the ship that measured 34 by 34 inches, (86 x 86 cm.), when fully opened out (and I know this for a fact for I've just now measured it).

I pointed to the middle of the ship on the side view, which turned out to be "B Deck" and from there to the middle on the overhead view, and that got my finger pointed at cabin B106. That was a two-berth, but I could have the one right next to it. So our cabin number was B108, and was situated just about dead centre of the ship, no matter whether it was viewed from the side or overhead.

I was probably the only person who was intent on being in the most stable position on the ship; maybe that is why that nice lady presented me with a floor-plan of all the decks, showing where everything was situated. Or maybe she just wanted to give me a memorable souvenir, for no one else that I spoke to on the journey got one. I have treasured it for all these many years, along with lots of other things (menus, canceled tickets for the family, a copy of the ship's log, daily entertainment schedules, etc.) regarding the journey. They are for my descendants.

~ * ~

At the same time as we were first talking of moving to Canada, there was another cousin of Mary's who was also interested. Her name was Sandra Bence and she was being courted by a young man by the name of George Whitelaw. They were continually dropping in on us to find out what we could tell them about Canada, borrowing brochures and—hey, *that* was who borrowed the book on Canada that never was returned to me. Yes, I know it was!

Neither Mary nor I thought too much about this with the two love birds, thinking it was just idle curiosity, until one day they dropped in to tell us that they, too, had made arrangements to emigrate to Canada and would be on the same ship as us. They had decided to get married and use the cruise to Canada as their honeymoon!!! This young lady, Sandra, had been a flower girl at our wedding in 1954 so we were quite delighted at the thought of their company. They were heading for Kingston, east of Toronto, to be met by George's uncle. All I was glad about later was that I hadn't influenced them in any way to make the move. I also wondered later if they would have come with us to Australia, if we had decided in favour of going there.

~ * ~

We were informed that our ocean liner would be sailing from Liverpool to the "Tail o' the Bank", the area close to Greenock, Scotland, on the River Clyde. (The Clyde is special to me as my favourite Scottish song to perform is "The Song of the Clyde," which includes the line, "From Glasgow to Greenock, with towns on each side, the hammer's ding-dong is the song of the Clyde.")[8]

---

8   Excerpt from "The Song of the Clyde," by R. Y. Bell and Ian Gourlay.

Greenock was where the Empress of England would anchor to take on the passengers from Scotland. They would be ferried from the pier out to the ship in a service boat (motor launch). I explained all of this to Mary and I could see her face changing as I told her. Quite suddenly she remarked, "Ah'm no' gaun'." (I'm not going.)

"Whit dae ye mean, 'ye're no' gaun'?"

"Jist that, Ah've changed ma mind and Ah'm no' gaun'."

"Ye canna jist change yer mind like that. Ye must hae a reason. All the arrangements are made and we hae oor sailing date."

"It dizny maiter; Ah'm no' gaun."

"Well, tell me whit made ye change yer mind like this at the last minute."

"Ah'm jist no' gaun' and that's that!" She had tears in her eyes and then she suddenly dashed upstairs to our bedroom and threw herself, face down, onto the bed. Of course, I had to follow. I lay beside her and put my arm around her.

"Mary m'dear, ye've got tae tell me whit it is that's botherin' ye, and then Ah'll no ask ye any mair."

"Ah'm jist no' climbin' up *they* rope things so Ah'm no' gaun'!"

"Whit dae ye mean, 'they rope things'? Ah've nae idea whit ye're takkin' aboot."

"Oh, *they* rope things ye see in the movies that's on the sides o' the big boats that the men use tae climb up tae the top." Then it dawned on me exactly what it was that scared her.

"Och Mary, ye're affa daft. They don't climb rope ladders tae get tae the ship, that's only whit the commandos and seamen do in war movies. When the wee boat takes us oot tae the big boat, there's a big stairway just like we hae in the hoose here that takes us up tae the deck."

"Ur ye sure? For Ah wid hate tae be stuck there!"

"Oh, Ah'm positive. Ah widna tell ye this if it wasna true. Ye've nothin' tae worry aboot. Ye'll be fine."

~ * ~

As it so happened she definitely didn't have anything to worry about, for when we arrived at St. Enoch's Station in Glasgow for the train to Greenock, we were told that because the weather was so bad, we had to take the train to *Liverpool* and board the ship there. (I remembered the time when I was just back from Egypt and on the train to Glasgow; I hadn't expected to ever be journeying back down that same route in the future—hah!)

The train platform was packed with emigrants and well-wishers—lots of laughter and lots of tears. Just minutes before the train moved off, the friends and relatives who were giving us the send-off started to sing, "Will Ye No Come Back Again?"

The words, "Will ye no come back again? Better lo'ed ye canna be"[9] mean, "Won't you come back to see us? Better loved you cannot be!" It's an old Scottish song, written for Bonnie Prince Charlie, but *now* it was being sung to us and I'll bet there wasn't a dry eye on the train. (*Never mind the train*, there isn't a dry eye here as I'm writing this, for it sure brings back some tear-jerking memories!)

We were told that the train wouldn't be leaving for Liverpool until early afternoon. The passengers didn't know that I was the one responsible for arranging our meal, for when I realized that there were no plans made to feed us, I went to see the station-master to get food arranged for everyone. So all of us who were to be on the train had a free meal in the station restaurant. I don't know whether it was Canadian Pacific (who owned the Empress of England) or British Rail who paid for it, but most likely it was C.P. Otherwise, we would have gone without or had to buy our own lunches. (Sometimes, I *do* come in handy!)

* ~ * ~ *

---

9   Excerpt from "Will Ye No Come Back Again?" by Lady Carolina Nairne (1766-1845).

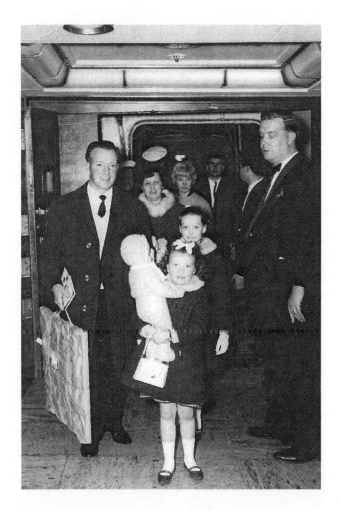

Off to Canada! Boarding the S.S. Empress of England, November 2, 1965 in Liverpool.

CHAPTER TEN
Atlantic Ocean Crossing Ends in Collision

Mary and I were still smoking cigarettes at that time. On the nonstop, fairly straightforward journey to Liverpool, we found that we had to ration ourselves; otherwise we would be out of smokes by the time we got there.

As we were climbing up the gangplank to board the ship, the ship's photographer was there to take photographs of the oncoming passengers. This was good, for I still have them to remind us of the voyage. (However, don't think they were free!)

While we were still living in Wishaw I had gone into Glasgow to arrange our cabin where I got a plan of the ship. Though I didn't really have to ask for directions to "B" deck, the staff was only too eager to assist anyway. We were shown where the elevators were and I managed from there. We also didn't have to bother dragging our carry-on luggage with us, which was terrific. We were told to go on board and everyone's luggage would be delivered to their cabins.

Our cabin had four berths (beds) in it, two on each side, one above the other. There was a little ladder hanging on the back of the cabin door to allow whoever was on the upper bunks to climb up there. On entering, Mary's bunk was on the lower left, mine was upper left, Shirley's was lower right and Audrey's was upper, as she was the eldest child and more capable.

Once we were settled on board ship, I made enquiries as to where I could purchase some cigarettes. I was told that I wouldn't be able to do so until the ship had cleared customs, which would be after we had set sail and were beyond the "three-mile limit" (extent of the territorial jurisdiction of the coastal country) and apparently this wouldn't be for some time yet.

About half an hour after this, the broadcast system announced that the ship had cleared customs and the lounge was open (though the ship was still tied up). I then suggested to Mary that we go there to have a wee drink and buy cigarettes. First we took Audrey and Shirley to the Children's Play Room where they were supervised.

Mary wanted a vodka and orange and I ordered a rum and coke, plus the packet of 20 cigarettes that we were rather anxious to have. The steward brought our order and I gave him a ten-shilling note to pay for it. He then gave me my change and I told him that he had given me too much, as it was not a pound-note I had given him.

He replied, "I know, Sir" and then told me the price of each item. I think if I remember correctly that the rum and coke was one shilling and sixpence, the vodka also, and the cigarettes were cheap. The whole bill came to somewhere around five shillings. If I had gotten these onshore, I would have had very little change, if any! Unbelievable! That was my first education on just how much duty and taxes bump up the price of things.

~ * ~

After we had replenished our supply of cigarettes, I escorted Mary back to the cabin, and then went to bring the girls to her. Then I went exploring (typical for me). As the ship moved away, it was like the movies with all the people waving. She was blowing her horn and the thrill of it all was immense. I stayed there for quite a while, moving from one side of the ship to the other, watching as she moved down the Mersey River on her way to Canada.

What a surprise I had when I noticed a ship tied up just a little bit away from us! On the stern was the name "Devonia." This was *too much* of a coincidence, for I'm sure *that* was the Devonshire troopship on which I sailed back from Egypt years earlier! I knew she had been privatized but I never thought I would ever see her again. Wow!

~ * ~

There was a lady steward assigned to look after a certain number of cabins. Our steward introduced herself to us shortly after we got settled in. She said that if there was any special service that we would like, she would certainly try to assist us.

"What do you mean by 'special services'?" I asked.

"Well, Sir, something like wishing to be wakened with a cup of tea and a couple of biscuits every morning before going to breakfast."

I could sure get used to that, so could my wife—we found it great! We certainly remembered to leave an envelope with some money in it for her at the end of the voyage.

~ * ~

The seas were very rough and the stabilizers were not being used as the ship was doing more pitching than rolling. As people on board were getting seasick, there was an announcement made to advise passengers that injections to combat this were available to anyone who wished one. Mary and the girls got fixed up but I figured that I had enough "sea-blood" in my veins that I didn't need one. It turned out that just about everyone on board became ill. The girls got queasy and Mary had to look after them quite a bit.

I went down for breakfast the third morning after we had set sail and found the dining room empty. I seemed to be the *only person* on board that wanted to eat! I was surrounded by stewards from the different tables all kidding me and saying they wanted to serve me. (Every table had its own steward for the whole journey). Mary could really have made it there but she elected to stay with the children who weren't feeling very good.

I thought afterwards that maybe there were others who had to do the same, maybe stay and watch over their children. That was the only time that happened, but all the other sittings were quite light, with *lots* of empty seats. The lightness of the seating wasn't due to the shortage of passengers, as the mealtimes were split into "sittings" to take care of all the passengers at different times.

~ * ~

There was a sort of "jutting-out-half-round area" on the uppermost deck at either side where the railing was, at about half way down the length of the ship. (I think it was for the pilot because, as it jutted out, it allowed a person to look along the side of the ship.) I was up there at different times throughout the day and sometimes around midnight, holding onto this half-round railing, facing the direction the ship was traveling and relishing the roll of the sea, the wind and the rain blowing on my face. Mary said I was "affa daft."

Though we were planning to live inland, in the Toronto area, we really should have settled down in some place where there was a dockside, with ships at anchor or tied up at the pier. I have always loved that sort of thing and I've always missed it, even when I lived in Motherwell and Wishaw. I guess I've got the sea in my blood!

I really figured that "my spot" on the uppermost deck would have been a great place to take all the folks who were seasick. I think it would have cured them in no time! While up there, I reminisced about wanting to go to sea when I first left school. However, my mother wouldn't allow it, as I had already lost at least two uncles (that I know of) by drowning. I can remember when I was about five, maybe six, that Uncle Charlie was leaving to go to the fishing boat he worked on. When saying our goodbyes, I asked him for a penny. He bent down beside me, mussed my hair with his fingers and told me that he would give me one when he came back. Alas, he never did. That was the last time I saw him.

I had taken "Navigation" when I was at Campbeltown Grammar School with the intention of going to sea. There I learned enough about plotting a course on charts, that by the time I had finished school, I was capable of laying a course from Liverpool to (say) Montreal. It wouldn't have been too difficult. I can still rhyme off all the compass points from memory, starting at North and going clockwise—North by east, North-northeast, Northeast by north, Northeast, Northeast by east, East-northeast, East by north, East, etc., naming every point, and ending up with North by west and then back to North again (32 points).

~ * ~

During one of the dinner times, a couple who sat at the next table to ours, and whom we had occasionally spoken to, sent a half bottle of wine to our table. They were Jack and Jean Pepper from Manchester. They had a little girl, Debbie, who was three. We became quite friendly and would sit with them during the evening while dancing in the Empress Room. Pretty soon it was as if we had been old friends for years.

(I don't know if I should tell this incident, but I will against my better judgment.) One evening, Mary and I went down to the Empress Room for a spot of dancing and ended up sitting at the same table as Jack and Jean. This wasn't the first time that Jean and I were able to get onto the floor together and eventually we ended up dancing quite close to each other, especially when we were on the opposite side of the floor from where our table was and our spouses couldn't see us! This time we were dancing very close to each other, cheek to cheek, breathing heavily into each other's ears. I found her pretty sexy and she seemed to think I was, too.

At that point Jean said to me, "Ian, will you come round to my cabin half an hour after we leave here and come into bed with me? I want to be with you so much!"

Well, I didn't know what to say for I didn't expect an invitation like that! "Jean," I said, "What about Jack, he'll be there."

"Oh, he's just about blotto now so he'll fall asleep as soon as he's in bed."

Well, I didn't go to her cabin—but I must admit that I wanted to. I was sorely tempted and would have liked to think that my Salvation Army training was keeping me from temptation, but I think it was more that I thought it would be extremely risky with Jack right there—blotto or not! I'm embarrassed to say that thoughts of Mary and the kids surely weren't upwards in my mind, which they should have been!

~ * ~

The ship's usual passenger capacity was 1500 but, as this was an extra, "last minute" sailing, there were only 500 on this voyage. There were only five first class passengers, among them the famous band leader Guy Lombardo. (Guy Lombardo and His Royal Canadians were especially famous for their New Year's Eve broadcasts across North America until his death in 1976 and became noted for their playing of Robert Burns' "Auld Lang Syne" at the stroke of midnight each December 31st.) This lack of passengers interfered with the buoyancy of the vessel as she was a lot lighter than normal. A thousand less people plus 25 thousand cubic feet less of cargo sure mounts up to a lot of missing weight!

We would go to the Empress Room every night and it was quite funny trying to dance. Oh, not the type of dancing that's done today—that would have been easy. I'm talking about dances like fox trots, quicksteps, modern waltzes—ballroom stuff that most people of my era enjoyed (and a lot fewer people know how to do today). We would be doing fine and then the ship would pitch or roll and send us out of control into the tables and people surrounding the dance floor. It was a lot of fun, too, watching the pairs of "super-expressive dancers" who would be doing their fancy stuff. At times the man would spin his partner around, then the ship would tilt and he would miss catching her. Sometimes both of them would land on the floor. This happened quite a few times but they took it with good nature and laughed it off.

A lot of people were seasick. An Edinburgh woman came on board ship from our train and went *straight to sickbay* where she remained for the *whole journey*. I don't know if it was psychological or what.

To make it worse, this lady had six or seven children who were at a table not too far from ours in the dining room. Without their mother present, their manners were atrocious! The father was already in Canada, and there's the possibility that she decided to let someone else look after them for the journey over the Atlantic. They were all pretty close in age. The oldest looked about eight and the youngest about three years old. They would throw fruit, yell things like "want mair tatties" (more potatoes) after they had used their spoons to catapult most of the mashed potatoes around the dining room. Each table had its own steward (waiter) who was responsible for attending to "his" passengers. All of us agreed that the steward in charge of those brats deserved a medal at the end of the trip!

Audrey, Shirley and I went to the indoor swimming pool one day to see what it was like. It was away down in the bowels of the ship and there was no one else there. We went to our separate changing rooms and I was first in the water. I dove in, came to the surface and suddenly panicked. When I surfaced after diving into the water, I happened to look to the side and saw that the water was at a great angle, lapping over the top of the side. The opposite side was showing the corner created by the wall and bottom of the pool. It took me a few moments to realize that I was the only one onboard ship that was level. I was in the water and the ship was leaning away over.

~ * ~

Every day, Mary and I went up to the promenade deck for morning tea and biscuits. (The daily schedule said "bullion or coffee" but we were delighted that there was also tea. It was always very difficult to get Mary to muster enough courage to do this as she was mostly terrified on board the ship. I think she would rather have left everything behind and flown *over* the Atlantic instead of sailing *on* it.

Anyway, I got Mary seated, then went to order for us and returned to our table with the two cups of tea. Then I went back, got our biscuits, returned to our little table and sat down. As soon as I did, the ship gave a big roll and my cup and saucer slid right off the table to dump the contents over my lap. I had two pairs of trousers in my carry-on luggage so I went to our cabin and changed, gave a request to our steward to have my pants dry-cleaned and then I returned to the promenade deck to join Mary. I got some more tea and a little later returned to our cabin.

It was getting close to dinner time (later we'd learn to call it "supper time") so Mary suggested that I go up to the children's playroom to get our daughters. I should point out that, as there were no *first-class* children on board, the tourist-class playroom was converted into a photographer's studio and any children on board were allowed to use the first-class playroom. (I think that it is very difficult to find "first-class children" *anywhere* now-a-days! Joke!)

To get there I had to do a lot of walking and climbing of stairs. Closing off the perimeter of the ship's decks were walls or "bulwarks" with wooden double doors in the middle of them. These doors didn't go all the way to the floor; they stopped about a foot short, meaning that anyone entering had to raise one leg at a time to get over the bulwarks. The reason for this was so that sea-water was prevented from splashing into the area during heavy seas, keeping that area dry. Anyway, I had to go through two sets of these doors to get to the first-class children's play room.

Keep in mind that as the seas were continuing to be very rough, the extent to which people were getting seasick seemed to worsen. As I stepped through the second set of doors, I hadn't noticed that someone had vomited just inside that doorway. I slipped and landed right in the middle of it. *Yuch!* That was my *second* pair of trousers that day ... but hey, it wasn't really too bad, as this was "first class" seasickness I was covered in; so it must have been of better quality than just ordinary tourist-class seasickness! Maybe it was even Guy Lombardo's and, seeing that I was a bit of a musician, it should really have been all right!

I didn't have any more trousers at hand. I had others but they were down in the hold. I had to rush back to the cabin after I got the girls and then try to find out if I could make a swap for the ones that I had handed in earlier that afternoon and then get the "tea-spilled-on" ones cleaned later. My pants couldn't be located on short notice, so I had no choice but to ask if I would be allowed down in the hold to look for my other suitcase.

"Yes certainly, Mr. Morrans, we will do all we can." There was *nothing* that was too much trouble for the people of Canadian Pacific!

I was escorted down, down, down, down. I lost count until we were in the very bowels of the ship. If there had been the usual number of passengers, I would never have found the case I needed, but I *did*, partly because the hold was quite empty, partly because we Scots were last to board. I think a whole lot of luck came into play, too. Rather than take a pair of pants out and close it up again, I just took the case with me to the cabin and kept it there for the remainder of the voyage.

~ * ~

There was always something to do on board the ship. If you wanted to be lazy and do nothing, then that was up to you. But from shortly after breakfast, the people in charge of entertainment made sure that if anyone was looking for something to do, there was a folder called "Entertainment" placed at each cabin door the previous evening, informing the occupants of the time and place of the following day's happenings.

What caught my eye and started my "creative juices" working was a children's fancy dress parade with prizes for the best costumes. After a bit of thought, I decided that a good character for Shirley would be the old lady from "Mary Poppins" who fed the birds. So I scrounged some cardboard and silvery wire from the crew and set about "constructing." I then fashioned a tray to hang from Shirley's neck with string. In this tray was a bunch of small "puffed-up" brown paper bags (which again, I got from the crew, who were only too willing to help out anyway they could). I made a sign for the tray, "Feed the Birds, tuppence a bag" (proper word is "twopence"). Next I fashioned a pair of glasses out of the silvery wire. With the tray, glasses and some make-up, Shirley looked just right!

We were thoroughly delighted when she won first place! Funny thing though, when she was doing a little shopping in the ship's gift shop earlier that day, Shirley had her eye on a nice little sailor doll; so we got it for her. And wouldn't you know—that was the prize that she won! The very same, which meant that then she had two identical sailor dolls. When this was explained to the person in charge, he said, "No problem!" Shirley turned in the sailor doll and was then allowed to pick her own first prize! She picked a lovely little Scottish doll that she still has to this day! Now, how's *that* for "Public Relations?"

Audrey was dressed as a "Balloon Lady" and as luck would have it, she came in third. We've racked our brains and none of us, not even Audrey, can remember what kind of prize she won.

~ * ~

We would see the newlyweds, George and Sandra (Mary's cousin) quite often, have tea with them sometimes, but mostly we left the young lovers to look after themselves. However, one evening they were visiting our cabin and it must have been later than usual, for the girls were going to bed. I took the little wooden ladder off the door to clip it onto Audrey's bed so she could climb up. It was then that George remarked, "Oh, *that's* what that ladder's for!"

Did we ever laugh! There were only two of them so they had a two-berth cabin, one berth (bed) above the other. When I teasingly chastised him for not using the top berth, I don't think I ever saw anyone so embarrassed! This was almost when we were about due to get off the ship, so they both had used only the lower bunk the whole journey. Their quarters might have been a little cramped; but, well, why not?! After all, they were on their honeymoon.

~ * ~

About this time I had to go to the Purser's Office to change my British pounds to Canadian dollars. I had been educating Mary on the new currency as best I could. It wasn't a new system to me, for the Egyptian pound also worked on the decimal system. When I had all of our British currency changed, I went down to the cabin to give Mary the money and also to give her a better idea of the coinage, now that I actually had some. I started counting out what I had.

"Here, this is a 20 dollar bill, here's a 10; so Ah'll count it out for ye." I got to $180 and I had nothing left! I counted it again; then again. I was short $100. I then went straight back to the Purser's Office and explained that I was $100 short on my exchange. I was told that they couldn't do a thing until they had closed the office and balanced their books that evening.

When I went back to the cabin Mary asked me if we would get our money and I told her, "Aye, we'll get oor money, but only if they havna given a hundred dollars extra tae someone else."

Imagine our relief when I was paged the following morning to go to the Purser's Office. The hundred dollars was waiting for me, and it was *very* important, as it amounted to more than a third of our "worldly wealth!"

~ * ~

Most of the Atlantic was behind us and it seemed as if we had it beat. We were fast approaching the uninhabited Belle Isle, off the coast of Labrador. I was a little more persistent in trying to coax Mary to come up on deck, telling her that we were getting quite close to land. She still wouldn't go up to the top deck.

Belle Isle came and went. I went up to the Promenade Deck to look at the chart that indicated each day's journey to find out just how far we were into the Gulf of St. Lawrence. I found that we were in the Strait of Belle Isle and that there was land not too far from either side of the ship. I went down to the cabin.

"Mary, ye've got tae come oop tae the deck and get yer first view of Canada. There's nothing tae worry aboot noo. Ye can see land on both sides, so come on oop."

It took a little more coaxing, but she did put her coat on and came with me to the top deck. As we were looking at a chart, I explained to her how it indicated each day's progress on the ship's journey, showing her where we currently were and explaining just how safe we were.

Standing close to us was an old man that I had spoken to a few times over the extent of the voyage. He had told me that he had made "so many" crossings of the Atlantic by sea. While we were looking at the chart he came over to us and said something like, "I see you've managed to get your Missus up to the chart to show her where we are. Well, it was just about where we are right now, in 1947 that the S.S. 'Something-or-Other' went down and all hands were lost."

Mary was gone in a flash, straight down to our cabin at close to the speed of light! I turned to the old guy and angrily shouted at him, "DID YOU HAVE TO OPEN YOUR STUPID BIG MOUTH

AND MENTION SOMETHING LIKE THAT, YOU STUPID OLD BUGGER THAT YOU ARE? DON'T YOU HAVE A BRAIN IN YOUR STUPID HEAD? THE NEXT TIME SHE'LL BE UP HERE IS WHEN WE'RE GETTING OFF THIS BLOODY SHIP!"

We still had a day and a half's sailing time to get to Quebec City and we *would* have to meet that brainless, mouthy old idiot! I felt like throwing him overboard!

~ * ~

Finally it was the eighth day of November, 1965 and soon we would be putting our feet down on Canadian soil. The ship was being towed into what I think was called the King George V Berth. We were informed that we would be allowed to go to the city for four hours if we wished, but had to be back at a certain time (I think five o'clock).

Mary, Audrey, Shirley and I got off the ship (and believe me, it certainly was funny trying to walk on dry land after having got used to our "sea legs"). We got a taxi up to the city and did some shopping around the various main streets. Mary bought a couple of little cheap things. After we had been wandering around for quite a while and it was beginning to get dark, I noticed a café sign. In Scotland we could get a cup of tea at such a place. We went in and sat at one of the empty tables. As the place was quite quiet, we didn't have long to wait. The waitress came to our table and said, "Oui, Monsieur?" I didn't know what she was saying, but I responded, "Two teas, please."

"Pardon?"

"Two teas, please."

Again she repeated in a slightly higher voice, "Pardon, Monsieur?"

I did my best to make her understand exactly what I wanted by putting an index finger horizontal on top of the index finger of my other hand to form a letter "T", then indicating someone drinking, then repeating, "Two teas, please." I got the same response again and again, with "Parle vous Francais?" added.

Finally I said to Mary, "She doesn't know what we want, so let's go," and she agreed. I thought it was terrible that here we were in Canada and this poor soul couldn't speak a word of English. Remember . . . this was in *November 1965!*

We all went outside and it was already dark and snowing. I can honestly say that from that day to this I've never seen such big snow flakes; they were monstrous—about the size of a dollar coin. It was snowing so heavily that it was difficult to see across the street. We all agreed that we should head back to the ship.

I don't know how the taxi driver managed to see his way, but he did manage to get us back there. And hey, *he* could speak English—especially when I asked him how much we owed for the ride!

Apart from letting off the people who were going to Quebec, the reason for the long wait at Quebec City was to allow the ship to take on provisions for the return journey from Montreal back to England. Why this couldn't have been done at Montreal is still a mystery to me. The only reason I can think of is, since Quebec City is the capital of the Province of Quebec, maybe politicians swung things for their constituents to profit as much as possible so that they would vote for them at elections! Stranger things have happened.

~ * ~

We had dinner, returned to our cabin and then, about six o'clock in the evening, felt the ship leaving the berth and being towed by tugs out to the middle of the St. Lawrence, ready to head upstream to Montreal. The engines hadn't been stopped, for we could always feel a very slight vibration going through the whole ship. We were in our cabin and as far as I knew, it was still snowing.

This was the last leg of our journey and the next day we would be in Montreal where we would disembark from the ship to go by train to Pickering, near Toronto in the Province of

Ontario. I felt that the journey right from the start in Glasgow was filled with unpleasant things happening to *me*. First off, there was the ship not leaving Liverpool to pick us up in Scotland, then having to arrange a meal for everyone at the station, then having to travel to Liverpool instead of Greenock, bad weather all the way across the Atlantic, getting tea spilled all over me, falling into vomit, having to spend a considerable amount of time hunting for my suitcase, not being able to get a cup of tea in our supposedly English-speaking new country and a few other things which were more an inconvenience than anything else. Though the journey was nearly over, I was disgusted by it all. However, soon we would be getting settled into a new house and all would be well—wouldn't it? (*Well—wouldn't it? ... Please ...?*)

~ * ~

Suddenly there was a very slight "bump"—so slight that it was hardly noticeable. (Talking to Jean Pepper later, she said that she only noticed it because she was in the process of washing her face and became a little unbalanced as she was bent over at the wash basin at the time.)

Everything went silent. The ship wasn't moving.

Nothing happened for ages. The ship just sat where she was. Time progressed and still the ship didn't move. A few people out in the hallways asked each other if anyone knew what was happening. Nobody did. I went back to the cabin and spoke to Mary.

"Okay, M'dear, Ah'll go oop, try tae find oot what's gaun on and then come back and let ye know as soon as Ah learn anything." I didn't get very far when I met another passenger who told me that we had been in a collision with a tanker.

'Bloody wonderful!' I thought. 'It's not over yet; it seems like we were not meant to move to Canada. All these lousy things are happening. I wish we had stayed in Scotland.'

I went back to our cabin and told Mary all that I knew and that I would go up to the top deck and find out more. I can remember her asking me if the ship was sinking and I told her that I didn't think it was anything as serious as that as the ship's horns were silent.

"Oh Ian, stai here wi' me and the lassies. What if the ship sinks and ye're not here? If ye're here Ah'll know where ye are but if ye go tae find oot what's happening, we might no' see ye again. Please stai here."

"Mary m'dear, there's nothing tae worry aboot. This is a whole different situation from the Titanic. Ye'll be alright, Ah'm just going oop to find out what's happening and Ah'll come straight back doon again. Ye know it was only a slight bump."

Well ... there *was* a lot of activity around the front of the ship and very slowly we were getting towed back to the quayside by the tugs that had pulled us out. We had been sitting motionless for two solid hours.

Jack and Jean Pepper came to our cabin. Then Jack and I decided that we should go up and have another look, as the ship was now tied up alongside the dock. There was another chap that Jack knew (I had spoken to him a couple of times) and the three of us went off the ship to get a better look at what had happened.

We could hardly believe what we saw! A fantastic amount of activity was going on as we walked along the side of the dock towards the bow of the ship. Lots of floodlights were set up and many people congregated. It boggled our minds to see the bow of the ship *burst wide open*. There was a hole that I could have walked through with my arms held wide if I had been able to walk on water, and I wouldn't have been able to touch the sides or the top. It looked like there was the same amount of damage *below* the water-line. *And all we had felt was a very slight bump*! You could swear blind that the front of the ship had been opened up by a giant with a great big can opener! It's a good job that there were watertight doors between each section, or we'd have been in the life-boats and Mary's fears would have been vindicated!

By this time it was really dark and close to nine o'clock in the evening. The snow had gone off before we were docked, but it was probably the snow that was responsible for the accident as it probably "blinded" the radar. That's my theory, at least.

It was probably Jack who suggested we go for a beer if there was a place nearby. I know it wasn't me, for beer was the last thing on my mind. We were directed to a "tavern." We asked what a "tavern" was—had never heard the word before—and were told it was a place where you could buy beer.

We must have crossed a million railway lines to get to it, and this was through over four inches of snow in street shoes. When we eventually found the tavern, we wondered if we should go in. What a dump! The smell reminded me of the close at Steele Street in Glasgow. All I could smell outside was "boozy" pee!

This was the sort of place, back in those (good?) old days, which *only men* were allowed to enter. (We found the Canadian drinking laws absolutely crazy when we first arrived.) However, I had to agree with this men-only rule in this case, as no self-respecting female would ever want to associate with such a dump anyway. I know there was no way I would ever have asked or even allowed my wife to accompany me into such a place. But, *we* entered, more for the excuse of having our first beer in Canada.

I'm sure this tavern wasn't much bigger than 25 feet by 25 feet. There was only *one* empty table at the centre of the place, all the other tables were situated along the walls. We went in and sat down. As soon as we did so, everybody else at the other tables stopped talking and stared at us. Jack went to the bar and ordered three beers. As we drank, our conversation was mostly about the hole in the ship. The other people seemed content to sit in silence and watch us.

After we had finished our first drinks, the other guy in our threesome went for three more beers while the customers remained silent, still staring. We drank up and then I went to the bar and got three more. We drank them and the locals were *still* silent. There was not *one word* spoken by the people in that place while the three of us were in there—probably the most uncomfortable hour of my life. There was a possibility that those were the people I'd probably be "cussing out" the next day.

After we returned to the cabin, we were informed that the ship was not going to proceed to Montreal and the remainder of the journey would have to be completed the next day by train.

~ * ~

I'd rather not have witnessed the ninth day of November, 1965! The ship had to be unloaded and the Quebec longshoremen who did so treated our belongings as though they were worthless. While unloading our boxes, they slung their hoists into the holds, pulled the loaded nets over to the quay side, lowered them a little and then released two corners, allowing our possessions to crash to the ground from a height of about 15 feet. We, the passengers, watched all of this going on in broad daylight. I was almost sick thinking that all of our wedding china and the lovely egg-shell china that I had brought back from Egypt would be smashed to bits.

The organization was ridiculous; I felt as if an air raid had just taken place! Those ignorant Québécois had no respect for anyone's property whatsoever—they were bloody, *stupid* animals in my opinion! Decades later I'm still mad at them, for they had no consideration for the property of others. I couldn't understand those actions at the time, but I've since decided that it was because we were all supposed Anglo-Saxons. We were newcomers to a country whose people destroyed other peoples' hard earned property because we were, well, *newcomers*, and probably *didn't speak French*!

After the longshoremen had "unloaded" the ship (so to speak), we were allowed to go ashore and see to our boxes and things. [Sorry, I have to stop for a minute or so for I'm still all upset by those actions all those years ago! - - - - - - - - - - I went upstairs and made myself a cup of tea and spoke to my wife while I was up there. She asked me why I was making tea so late, for she knows that I try to limit my fluid intake in the evenings, otherwise I'm up at the bathroom in the middle of the night. I told her and she replied, "It's all water under the bridge, Ian." I know it is, but nothing annoys me more than gross stupidity, and even thinking about that particular episode sets me off!]

During the course of gathering all of our boxes together in one place, I had to go *personally* and find all of my boxes. They had been marked with a large sticker bearing an "M" for our initial. I've got no idea what the dock workers were doing; probably they'd gone for a liquid lunch when they should have been getting our stuff for us. It surely wasn't our responsibility to gather our boxes together!

I came across some sights that could make you weep. One of the most disturbing things was the sight of what was once a beautiful Telefunken Stereo cabinet that was smashed to smithereens—only fit for firewood! What a shame! What those idiotic longshoremen maybe didn't realize was that doing those kinds of things just causes hate, *lasting hate*, and most likely results in reprisals later in some way or another.

I'll never forget—or forgive—those bloody dock workers! When we finally opened our boxes, we discovered that almost *all* of our wedding china had been smashed. This was stuff that was supposed to be handed down to granddaughter and great-granddaughter. Not much of it survived, just a couple of cups and saucers. Funny enough, the Oriental eggshell china faired a little better. We have some cups and saucers, the tea pot and a few other pieces. As it was the most fragile, I'm surprised it wasn't completely destroyed. Lots of other things were, so please don't look for any forgiveness—EVER—in my heart toward those inconsiderate B#$&@$%&'s (saves swearing, doesn't it?)

~ * ~

A short time after coming to Canada, while I was relating the story of the restaurant and the attempt to get some tea, I said that it was such a pity that the waitress couldn't speak English. I was told that she most likely *could* speak excellent English and knew exactly what I was wanting, and if I had made an *attempt* to speak French, I would likely have got my tea; for tea in French is so close that it is almost the same! I couldn't believe that something so ridiculous could happen in a country like Canada!

Much later as I pondered about the waitress, the stony silence of the grossly ignorant customers in the drinking hovel and the destructive longshoremen, I wondered if I had done the right thing in coming to Canada. It wasn't very much later that the doubt got much deeper. I did vow that I would *never ever* visit Quebec, and you can put money on that! How could a person feel comfortable in a place that caused so much ill-feeling? (Off and on, I kept thinking, 'How ya doin', Australia?')

~ * ~

A train was laid on to take us immigrants to Montreal. We had met up with Jack and Jean Pepper again and were all in the same compartment as the train left Quebec City. It wasn't too long a journey; I think about five hours. During the journey, I asked if anyone wanted something to eat as there was a dining car. I could go and get something for each of us.

Not wanting to appear selfish, I asked Jack and Jean if they wanted something and they said "yes." I can remember what it was that I got on the train. I also know that it was a long way from our coach to the dining car, as trains in Britain were nowhere near as long as this one; it seemed to stretch for miles. What I got in the way of something to eat was ridiculous for a dining car—a few half-stale sandwiches and a few cans of warm pop. What's more, I had to *pay* for them.

We should really have been dining in elegance on board the Empress of England if it had been able to continue to Montreal, and here we were, getting shuttled by a train that didn't have any decent food on board, and the ticket I paid for was supposed to take us to Montreal, *food included*. I've been singing the praises of Canadian Pacific while I have been relating the journey from Britain to Canada, but now I have to change my tune. There should have been decent food for us aboard that train, or at least a refund for the food that we didn't get. It was already paid for in our tickets. Then again, maybe Canadian Pacific had no control over the situation for there is always the possibility that proper arrangements had been made but not carried out!!!!

~ * ~

By evening we were approaching the outskirts of Montreal and the train was passing an industrial area. That gave Jack and me a look at some of the places that the train was just creeping by, as the interiors of the buildings were lit up. We passed a couple of machine shops and what we saw opened our eyes. (Jack was a machinist also.) Everything was so antiquated! All the lathes were belt-driven from an overhead shaft, and Britain had stopped using something like that practically when Noah landed the ark! We looked at each other in amazement and disgust and wondered what we had come to.

When we finally arrived at Montreal we were taken by bus from one station to another to get the train to Toronto.

Jack and Jean were still with us, but were told that when the time came to board the train, they had to get onto a different car as they were continuing on to Toronto. There was something like a four-hour wait in whatever station we were in. We really should have been taken to a restaurant and fed by Canadian Pacific, seeing that there was such a wait. It wasn't bad for us, but there were people there with infants who would have had it worse.

We were all hungry again and I went to find somewhere to buy food. I think it was quite late in the evening, and there wasn't much to pick from. I ended up with bags of French fries and some more pop for everyone, including the Peppers. I was left to foot the bill as Jack didn't offer to pay for his share of their food—neither the sandwiches and pop on the train nor the "food" I had just bought in the station. However, I had offered to buy and that was it, so it wasn't a big deal. It was, I think, about two years later that the situation was explained to me, so I'm shooting forward to that time for a brief moment, otherwise I may forget to mention this.

We had kept in touch over the years, visiting each other now and then and were visiting their rented basement in one of the big houses on St. Claire Avenue in Toronto one Saturday evening. That was when Jack confided to me that he came off the ship with only 18 dollars in his pocket! That was *all* they had in the world. And I had thought *we* were bad off with just a little less than three hundred! That evening, he was very apologetic about not being able to pay his way that night a couple of years earlier, and I told him that I hadn't thought much about it, to forget it! If I *had* been aware of the situation, I would have insisted on, at the very least, getting some food for their three-year-old girl!

In the many years since he told me this, I've often wondered about his wisdom in sending a half-bottle of wine to our table on the second day of our journey, especially as we were complete strangers. It was certainly a very nice gesture, but in my book, one doesn't jeopardize one's family to make nice gestures. Also, they (Jack and Jean) were in the lounge every night during the voyage and although the drinks were cheap, they certainly weren't free! Also, the trip to the tavern was Jack's idea. I was trying to be very careful in our spending for I knew that we didn't have very much money to throw around. Now I realized the Pepper's had had a whole lot less!

Anyway, back to the morning of the 10th day of November 1965. It was daylight and the train was heading towards Toronto. Next stop would be ours—Pickering, Ajax (or Ajax, Pickering), whichever way is correct, I'm not sure. I had phoned Mary's cousin, May (collect) from Montreal to let her know what time our train was scheduled to arrive at their station. As the train slowed down, I hoped that May and Bill would be there to meet us as promised when they had first encouraged us to move to Canada. With the amount of misfortunes that had been happening in our lives since trying to leave Scotland, it wouldn't have surprised me one bit if there had been no one there!

* ~ * ~ *

Trying to make the best of it when we lived in Teoli's basement apartment in Toronto, Ontario, 1966.

CHAPTER ELEVEN

Oh My, What have I Done?

There was less snow in Pickering than we had seen in Quebec, just a little at the edges of fences and roads. It was dark when we had arrived and also when we left Montreal, so I had no idea what amount of snow to expect. Of course, I didn't know the geography of Canada at that time, completely unaware that Toronto was a good deal further south latitudinally than Quebec City.

As the train drew to a halt at Pickering Station, we had gathered ourselves together. This was the first step into what was to become our new home area. We were ready to hop off the train whenever it stopped. I opened the door, stepped out and there were only two people waiting on the platform—May and Bill.

When the train drew away we saw Jack and Jean, just a few coaches from us, waving goodbye. There was no reason why we couldn't have been in the same compartment for the journey from Montreal.

~ * ~

There were hugs and kisses all around as May and Bill met Audrey and Shirley for the first time. Bill and I shook hands; then it was off to their car—the guys carrying the luggage while the females followed behind. Big relief—we had made it! All of our boxes were to be delivered later to Bill's house.

Bill and May's house at 879 Grenoble Boulevard was everything they had said it was. I had never been in such a nice home in my life, even their half-basement was done up nicely and just like a part of the house. (I found out later it was termed a "finished" basement.) It was almost brand new and was what was classed as a "back-split." May and Bill really made us feel welcome and invited us to make ourselves at home. They had a nice piano that neither of them could play, although May had started to take lessons. When I said that I could play a bit, although it was only by ear, they were both delighted.

May's younger brother, Jim, was the person who told May to tell me to get to Canada fast as there was a job waiting for me. He and his wife Helen lived about five minutes' walk away. May told us that they intended to visit during the evening to meet us. Mary already knew both of them previously when they had lived in Scotland, and it was like a reunion for her, giving them all the latest news. I met the two of them for the first time when they came round to May's place that evening. We got along just fine.

A Singsong with friends. I'm at the piano.

Later that evening, after I had had a couple of beers, I sat at the piano and we actually did manage to have a bit of a singsong. Jim mentioned the job that was waiting for me and said that he would make enquiries when he was back on day shift.

The next day was Thursday, November 11th, Remembrance Day. May took us to a few large stores to let us see what Canadian stores were like. Remembrance Day wasn't fully recognized as a day of closing like it became in later years and we were in one of the large retail outlets when the one minute's silence at 11 a.m. was announced.

Quite a bit later that day, we were back at May's when she offered to drive us around on the weekend to let us see what the area was like and maybe get an idea of where we would want to live. When she mentioned rental apartments, I said that we wouldn't bother with anything like that as we were interested in buying a new house.

Quite surprised, May responded, "Oh that's very nice; then you came here with quite a bit of money, did you?"

"No," I replied, probably smiling, "I've never been in a position of having 'quite a bit of money,' but the Ontario government official in Glasgow told us that we just needed $200 to get into a brand-new house. If I start work right away, I'll have that amount of money in two weeks. Isn't that good?"

I can still see May's face so clearly in my mind's eye, taking on a very sad look. "Oh, Ian, why on earth would he have told you that? You need at least $4000 to put down to buy a new house."

"Well, that's what the book I read about Canadian immigration says, too; and although it doesn't mention the amount of money, it does say that nearly everyone in Canada owns their own home. That was one thing I made sure I asked about when I had my appointment in Glasgow. I even showed him the book. That was when he said that all I needed was $200; that I could buy something far better than I could ever buy in Scotland. It was the major factor in our decision to come to Canada."

It was now Bill's turn. "Ian, it's more than 40 years since one could buy an *old* house with $200 down. You can maybe buy an older, run-down house for $2000 down, but the days that guy was talking about are long since gone. I wonder what would make him tell you something like that."

'*So,*' I thought to myself, '*it isn't finished yet. I wonder what will be next.*' A few days later I found out.

In the meantime, Bill and I went down to the local beer store. We walked through the doorway and Bill was in front. I was intent in getting the door closed for it was cold outside. Bill called to me, "Ian, you have to get off the mat at the door before it will close." I'd never seen the automatic door openers before as we didn't need them in Scotland. I think everybody in the beer store was looking at me, making me feel very stupid.

~ * ~

Jim had seen his boss on the Wednesday, told him that I had arrived in Canada and had asked him about the job that was supposed to be there for me.

You've probably already guessed that, by the way things had been going for me up 'til then, most likely there would be no job waiting for me. You're right! When Jim told me this, explaining that his company had decided to cut back on hiring, I wasn't really surprised, although I was very disappointed. It would have been nice to have been able to start work without having to look around for a job, especially as I had no means of "getting around." I had hoped that things would soon be going my way. Jim said that his days off were Friday and Saturday and offered to take me around on the Friday to see if I could get a job.

It was around this time that I discovered there had been a two-day power outage all down the eastern seaboard from and including Montreal and Toronto to down around Virginia in the U.S.A. Well, all I can say here is that it was just as well that it happened *before* we had settled into Canada, for if it had happened two days *after* we had arrived, I think I would have *swum* back to Scotland, right then and there! And I would have been yelling all the way back, "Okay, that's it, I've had enough, *I have had it!*"

Things did brighten up a bit on the Friday. Jim took me to various places that morning and at one place I was rewarded by getting hired with a definite starting date. My first job in Canada was to be at Link Belt in Scarborough, a large company. I was to start work on Monday, 15th of November at 8:00 a.m.

The rate of pay was $2 an hour and it seemed to be around the standard rate for that time. Not great, but not bad. The job was assembling conveyor systems, mostly rollers. It wasn't a skilled job, but at least it was a job. I was part of an assembly line. My job included putting a bearing and a seal into a roller, driving a shaft into it, then installing a bearing and seal on the other end. You didn't need any brains to do that job, just strong arms to wield the heavy lead mallet that was used to drive in the shaft, so that the end didn't get burred over, which would have happened if a steel hammer had been used. I figured it would have been a good job for a chimpanzee! Requiring about the same level of intelligence, too!

My arms ached continuously. I added a step by freeing the bearing to make sure that everything was all right. (This bit I wasn't supposed to do, but no one told me this. I figured that it should be "right" before I passed it along the line.) During the first two weeks I was getting a ride into work with a friend of Bill's and I paid him some money for gas.

~ * ~

Have you got the feeling that everything was going too smoothly again? Well, if so, you're still ahead of where I was, for I was shocked to be fired after three weeks and two days! We had been working every Tuesday and Thursday night until eight o'clock. Don, the foreman, came up to me on the Tuesday about four o'clock and said that I would be finishing at 4:30. I innocently asked if the overtime was finished, and he said "No. You don't understand what I mean. What I'm telling you is that you are out the door at 4:30 for good. You are being *fired*!"

When I asked Don why I was being fired, his answer was (I can still hardly believe it!), "You are too particular with your work, and that's the trouble with you Brits."

I asked him to explain and he replied that I was doing things at the job that I wasn't supposed to do, so I was taking too long. What I had been doing apparently, was someone else's job as well as my own, meaning that there was a guy further down the line that was having a really easy time of it! When I told Don that no one had explained this to me and that from now on I would do just what I was supposed to do, he replied that the decision had been made and that was it.

As it was getting close to 4:30, I went up to Don and told him that my pay had better be made up to date. He said that I would receive it in the mail. My reply to him was that if they were letting me go, then it was their responsibility to have my pay ready for me and that I demanded it. That created a bit of a flap, and someone had to work late to get my pay cheque ready.

At 4:30, Don came up to me, wanting to shake my hand and wish me the best of luck. I told him to get lost. He then said that I was just being childish by wanting to get my pay right away. My reply to him was that I had never been fired from a job in my life, so if I was no longer of any use to them then they were no longer of any use to me and that whatever was due to me I wanted right now. (I could be a rebel sometimes, when it suited me.) I also found that it is just about impossible to get *any* kind of a job so close to Christmas—nobody left their jobs at that time. Plus that, there was nothing in my immediate area, I didn't know the surroundings and I had no car.

~ * ~

(I have to throw in some comedy here. It concerns Mary's adopted brother Ken whom I haven't mentioned before—he was born to her Aunt Liz before she was married and was then adopted and raised by Mary's parents.) Anyway, Ken had driven up from Toledo, Ohio to see us while we were living with May and Bill. I think he had got a few days off for U.S. Thanksgiving which fell that year on November 25. I had got to know him, but just a little, while he was still in Glasgow and quickly found out that although he seemed all right, we didn't have anything in common. (He, his wife Betty and their two daughters had first settled in St. John, New Brunswick a couple of years earlier and a little later moved to the States.) He had brought his ten-pin bowling balls with him, so we all went bowling. At least, Bill and Ken bowled. All I managed to do was throw a ball *backwards*, as no one had told me that I had to find a ball that had my size finger holes in it. It seemed that I was getting really good at getting people to look funny at me!

~ * ~

During the time I was employed at Link Belt we had been looking for a place to live. We kept watching the newspapers as we thought that May and Bill were getting anxious for us to move out. May's brother Jim kept telling us that if we wanted a place of our own some day then we had to find somewhere cheap to rent so that we could save our money. I thought that was a joke after I compared what rents were *in Canada* to what I paid for the nice town house in Scotland. The cost was easily four times as much as what I was used to, and *that* was for an *old* apartment! How could I find somewhere *cheap* when "expensive" seemed to be the standard?

A few days after I had started work, we saw an ad in the paper that told us of a couple who were looking for someone to babysit their two little girls and that two furnished rooms and a kitchen were included, plus they would pay $20 a week. (It worked out to less than 40 cents an hour, but we had no idea what baby sitters were paid). This was in the Kew Gardens area, at the extreme (east) end of Queen Street East, Toronto.

We went to see the people and they seemed delighted with us. It was dark when May and Bill had driven us into Toronto to their address, but they didn't come into the house. (I thought that May's brother Jim would be pleased, for we *did* find some place cheap to rent, although it did look a bit of a dump in daylight, tall and narrow, semi-detached!)

When they showed us where we were to be living I almost told them to forget it. We had to ascend a very narrow staircase (one person wide) four floors up to get to our rooms and the kitchen and the bathroom we would use were one floor below us. Our two rooms were very much like what I had lived in as a boy in the *slum* at #8 Lorne Street, Campbeltown, with sloping walls. We were to be *living in an attic.* And I found out something else, it is really amazing what some people classify as "furnished"—it was a pile of junk!

I was almost in tears. My exact thoughts were, **'What the Hell Have I Done?'** (You have to read that *slowly* with lots of emphasis to get the correct feeling.) 'What have we come to? What have we been reduced to by moving to this bloody country? Where are all the good things that we were told were here in glorious Canada? How come we're to be living in a crumby attic, with furniture that I would have thrown out to the dump before we came here?'

I had never felt so low in spirits in my life, and I've been there and seen some! I wished with all my heart that we had never made the move—that we were still back in Scotland where life was a heck of a lot better than this! I suddenly realized that we had made a major mistake and there was not one thing I could do about it!

~ * ~

We moved into our "crummy attic" on Sunday, 28th of November while I was still employed at Link Belt. Mary took up her duties as baby sitter the following day. During the evening she would sometimes disappear into the other bedroom to have a good cry—and I often felt like joining her. She had done this a few times while we were living with May and Bill. I always had tried to console her, thinking of the good times that were still to come. I couldn't console her anymore. What was I to tell her? "Look Sweetheart, we have a lovely home now and everything will be all right from now on."????

No ... what I *could* tell my wife was, "Sorry Sweetheart, we have nothing. We've been reduced to less than where we were before I came out of the Royal Air Force more than *ten years ago*." It was hard to think that we had left a really nice home and a nice job in Scotland to come to this.

On the other hand, Audrey and Shirley, now attending school, were adjusting much better and already talking with a Canadian accent! It took the two of them just three weeks to lose their Scottish accents. Rats!

~ * ~

We had to endure, there was no way we could ever go back, precious few of our possessions had survived and we were next to being stone broke, so we had to make the best of it. It didn't make one bit of difference whether we were in Canada or in Australia in this condition. So much for May's argument when she was trying to encourage us to move to Canada. We could no more go back to Scotland from here than we could have done from Australia. It seemed like it still would have cost us about a million dollars!!! I wished to hell that people would mind their own damned business!

~ * ~

Back to the baby-sitting that Mary had to do. She looked after our landlord's children during the day for a total of about six or seven weeks. One afternoon, four weeks after we had moved in, one of the little ones slipped on the wet grass in the backyard while Mary was hanging out the laundry, the result being a slightly scratched cheek (not much bigger than a mosquito bite).

After her "employers" had put my wife through the third degree and told her that she wasn't fit to raise children, I really got into them and asked them who the hell they thought they were. I told them that she had done an excellent job with our own two children, that they could stick their lousy rooms where the sun doesn't shine and that we would start looking for some place else to live. Unfortunately I had opened my mouth too quickly as we really didn't have any money to be able to move at that point. It seemed that things were getting worse instead of better! It was getting very, very difficult to keep a pleasant demeanor.

~ * ~

Not long after we moved into the "attic in the sky" in early December, May and Bill came in to pick us up and take us to meet special friends of theirs. When we came downstairs to the entrance, Bill said to me, "Watch your footing out there, Ian; it's freezing rain."

I looked at him and said, "Yes, sure Bill, so it is." Then, taking one step out of the door, I immediately fell on my rear end. It was sheer ice. I had thought Bill was pulling my leg. We'd never heard of that stuff in Scotland. I stood there amazed as I watched kids *ice-skating* on the *street*. The ice was at least thick enough to skate on. If I had written home and told my friends about this, they wouldn't have believed me!

~ * ~

We quite rapidly had gone through the wee bit of cash we had saved when I was working, buying clothes for the girls and other necessities. After I had been fired from the job at Link Belt, I went to Canadian Immigration as I thought it was my duty to inform them where I was and how I was doing. I told one of the men that I was out of work and asked him what I should do.

He took me into the office, asked me to sit down, got out a card file box and started to go through it. (Keep in mind; this was long before computers came onto the scene.) He had asked me what kind of work I was looking for. I told him that I was a machine fitter. His reply was that there weren't any openings in that field and was there anything else that I could do.

"I have a wife and two children to care for and will take anything to earn some money."

He pulled a card out of the box and asked me, "Do you have a driving license?"

I replied that I had (a British one).

"Let me see now," he said while reading the card. "Yes, you *are white* (they sounded just *slightly* more racist in those days), you speak English and have a British education. Yes, you will be all right for this job. Have you ever thought of being a salesman?"

I said. "No, but I'm willing to give it a go if there is a job for me."

The address given to me was for a Wonder Bread depot. I can't remember the street address but it was somewhere near Woodbine race track. When I found the place, I was offered a job driving a baker's van around Markham, selling bread and cakes door to door. The rate of pay was a dollar an hour and I had to work six days/sixty hours per week. (The year was 1965 but it could just as well have been 1925! Things were looking worse instead of better!) Anyway, I took the job. Better to be earning something than nothing at all.

~ * ~

I had one day's instruction about the Wonder Bread job and the best bit of advice I got from the guy who was showing me the ropes was, "Whatever you do, don't screw the customers."

My reply was that I didn't believe in doing *anyone* out of money.

"When I say don't *screw* the customers I mean don't go to bed with any of them. You will get invited, but don't do it." I thought he was kidding. He wasn't—I had offers from females even as young as 16! (Obviously, I hadn't learned North American jargon yet! There were many expressions that were new to us and also some that had entirely different meanings. Back home in Britain, "screw," meant "steal from;" "knock you up" meant to rap on your door to give you a call early in the morning to get you out of bed; and you would *never* ask a female to go for a "ride" in your car. Instead, you would offer her a "lift"—a "ride" meant having sex!)

Another example: shortly after we arrived in Canada, Mary and I were at a dance in Toronto. A group of us were standing and chatting at the edge of the dance floor when I announced that I was going to the bar for a drink. When I returned, a young, good-looking woman put her arm through mine and I understood her to say, "I like the way you roll your arse!" (What we call "arse" is referred to as "ass" or rear end in North America.)

I hesitated a little and looked down at one buttock and then the other, wondering what it was I did with my "arse" that got her attention. It wasn't until I thoroughly thought about it that I realized that she was saying that she liked the way I rolled my "RRRRs!" I guess she enjoyed the Scottish accent. Boy, what a relief!

Back to Wonder Bread—Harry, an Englishman, was in charge, sort of a foreman. He told me that I had to keep the evening of Christmas Eve free as that was the time I had to go around to all my customers and deliver the Christmas puddings that they had previously ordered.

Meanwhile I had learned that an outfit called John T. Hepburn in northwest Metropolitan Toronto (not too far from Steele's Ave. and Albion) was looking for skilled men. I went and applied for work. As soon as I spoke to the foreman I thought I had it made—he was from Scotland! I ended up getting one of the stiffest job interviews I have *ever* had. He asked me just about every question under the sun. As it turned out, he offered me the job. The wages were $2:10 to start, with an increase after three months. (I thought maybe things would finally start to look up.)

~ * ~

Before quitting at Wonder Bread I went to the Immigration Department to advise them, as I thought that they were supposed to know of my whereabouts at all times. (Silly me!) This time I saw a different person.

"I just came by to tell you that I have found myself a better job. I'm quitting the bread delivery van job and I'm starting work as a machine fitter on the Monday after Boxing Day." This was on the Wednesday. (I had taken some time off work to see about the job and I can't remember exactly how I worked it.)

"What on earth were you doing driving a baker's van?" he asked.

"That was all there was. There weren't any jobs for my kind of work in your files."

"Like *hell* there weren't! Who told you that?" he asked, pulling a huge pile of cards about an inch (2.5 cm.) thick out of the box, "These are all *'fitters wanted'*." (I didn't see the "big picture" right then. It took a couple of months.)

~ * ~

Next thing of course, was to quit work at Wonder Bread. When I drove into the yard with my van on Thursday evening (the office wasn't open when leaving there in the morning), the first person I met was Harry. As soon as he saw me he reminded me, "Ian, don't forget that tomorrow night is the night that you deliver the Christmas pudding." Friday was the 24th.

"Well Harry, got to tell you;" I yelled over to him, "I'm on my way into the office to tell them that I'm quitting today. I start work back at my usual line of work on Monday, the day after Boxing Day at Hepburn's, at more than twice what I'm getting here, and for 40 hours a week. Anything over that is time and a half." I didn't even wait for his reaction or listen to what he was saying, for I wasn't the least bit interested in how the Christmas pudding was going to be delivered. Besides, Cousin May was having a Christmas Eve party at her place and I didn't want to miss that.

House party. I'm dancing with a neighbour.

Well, the party was just great. We met a lot of May and Bill's friends and all had a really good time. That was where I first saw the "Liverpool Shuffle" dance performed. Mary and I learned it very quickly. We did it for years afterwards, at public dances and at parties at our house and at friends' houses. We taught many others to do the Liverpool Shuffle, too. That was what is known today as "line dancing." Folks nowadays think it's great and we were doing it in *1965*!

~ * ~

We were still at Kew Gardens, a couple of weeks after the New Year 1966. Mary was still doing her baby-sitting and, since my "outburst," there had been no more static from her "employers." However, to get to work at my new job I had to leave home at five o'clock in the morning to catch the street car that took me on the long, long journey to start work at 7:30 am, changing street cars a couple of times. Kew Gardens was at the extreme southeast end of the city and my new job was in the extreme Northwest! It would have been difficult to find a job any further away from where I was living.

We had to find some place closer for us to live. We still didn't have any money to speak of as we seemed to spend it as fast as I got it, since we found it necessary to replace our possessions. We also had no furniture. When looking at the classified ads in the newspapers there was nothing in the way of furnished apartments. Most likely no one moved at that time of the year. So we then started looking at anything that was available.

~ * ~

Incidentally, I had met a fellow Scotsman who told me of the terrible time he and his wife had had trying to find a place to live when they had first arrived in the Toronto area. Why did they have a hard time? Because they had four children! Nobody wanted them! Remember the immigration guy telling me that Canada was really more interested in our *children* than us, as they were the future? Someone should have educated him and also should have told the apartment owners! Even on the banners which flew from apartment balconies I saw the wording, "Apartments Available, No Children." And immigration thought it was a *joy* for some people to come to this land! I liked it less and less every day and continued to wonder how "Down-Under" was doing!

* ~ * ~ *

Our first car in Canada, a 1959 Chevy Bel-Air. Price $360, 7 years' old, and so rusty that the rocker panels fell off within weeks.

CHAPTER TWELVE

We get "Established" at Last

There was a vacancy that was an unfurnished basement of a bungalow in Plewes Road, Downsview. I decided to call and inquire about it. The rent was $80 a month (this was four dollars short of a week's wages before taxes and very much more than what we paid for our nice town house in Scotland, including our carrying costs of heat and water, etc.) It was owned by a nice old Italian couple by the name of Teoli. If they had one, I think I paid their mortgage for them!

Mary, the girls and I went to see the apartment and it seemed to take us forever to get there. It was not great, but at least it was nice and clean, fairly modern, and we didn't have to climb a very narrow staircase, four floors up! It consisted of a kitchen which opened into a living room, and then there was one bedroom. I told them that I would like to rent their place but I didn't have enough money to pay the rent in advance like they wanted. Old Mr. Teoli had been around and was wise to the world, telling me that I didn't need any money, all I had to do was go to the Immigration Department and tell them that I had to have a place to rent, that I had found a place, that it was $80 a month in advance and that I had no money.

"You go see dem and dey give you duh money for you rent fo' sure," he insisted in his broken English. "I give you paper to tell dem it is my place."

He was right. When I went to the Immigration Department, I was told to sign a form after the clerk had filled it in, stating all that I had told him. Then I had to wait for a few minutes and soon I had a cheque made out for $80 in my name and Mr. Teoli's name! When I asked him how I could go about repaying the money, he said that I couldn't, that it was a "grant." I wondered just how many immigrants had "hit" that spot for cash, even when they didn't need it? I'll bet that every Italian immigrant that lived in "Little Italy" in Toronto knew of the cash cow that was waiting to be milked! Whether they used it or not could be another story, but knowing the nature of a lot of people, you can bet that it was abused.

It seemed, I found out much later, that the Canadian Government was responsible for our welfare until we were in Canada two full years, but I was never informed of this. (I would imagine that was the sort of thing that, *rightly*, would be best kept a secret.) Not that I would have used this for any gain; I preferred to make it on my own. They can shove their help, although if anyone would have been justified in taking advantage, then I think my family would have been close to the top of the line. It would have been nice to get the *Ontario* Government to recompense us for what we gave up to move here on their encouragement. The only difference was that it was the *Federal Government* that owned the cash cow, so there was no point of milking them.

So we soon had a place to move into, the only thing was that we didn't have any furniture. In retrospect, I should have gone to the "cash cow" and told them that I had no furniture; maybe they would have supplied that, too!

~ * ~

The ad in the paper said, "Three rooms of furniture for only $299." I gave them a phone call, found that they were open late and told them that I was on my way from Kew Gardens. I rolled up the section of the newspaper to take with me, told Mary where I was going and walked down to Queen Street for a streetcar. Knowing that Queen Street West started on the other side of Yonge Street, it seemed to take forever to get there. It was early evening, the middle of winter and dark so I had a hard time watching the street numbers through the streetcar window and slightly misjudged it. I got off about a block too early and there just happened to be another furniture store right there.

As they had an advertisement similar to the one in the paper, strung across the top of their window, I went over to see what they had to offer. Remember it was January, a little after Christmas and practically *no one* was spending any money—especially on furniture. I walked into the shop and two guys were sitting there just killing time, probably the owners. When I entered they suddenly came to life and gathered around me.

Now, follow this closely. I think I can relate it just about exactly word for word as it went along.

"Yes, Sir, how can we help you?"

"Oh, I see by your ad outside that you have three rooms of furniture for $299."

"Yessir—over here please. This is the bed chesterfield and over here are the coffee table and end tables—and if you come this way, I'll take you upstairs and let you see the kitchen set and the bedroom set."

I went upstairs, had a look and then went back downstairs again. "Well, that is very nice. Just give me a few minutes and I will go along and tell these people who are waiting for me (showing them the ad for the other store). Then I will come right back to see you." I was sincere at this point about going to tell the other store.

"Oh now, just a minute, Sir. Before you go, I'd like to show you *this* bed chesterfield. It is quite a bit better than the other one and I'm sure we can make a deal."

"Gee, that's great! Just let me go to see these guys, for I did phone them to tell them I was coming."

"Sir, that kitchen set you saw upstairs isn't that good. What do you say that we exchange it for this one? It is much better."

At this stage, I realized just how powerful this little bit of newspaper was that I was holding in my hand. They probably knew that if they let me get out of the shop they would never see me again. "Well, that's absolutely marvelous; just give me a few minutes and I will be right back."

"Sir, see those plates and cutlery that are on this nice table? We will include them in with your purchase."

"Great, just great. I should be back in less than five minutes. Gee, my wife sure will be pleased."

"Oh—just a minute before you go—look over here, Sir; we will also include this floor lamp." (It was what we Scots called a "standard lamp", and the bedside table lamps were already part of the deal.)

"Hey, I can't believe it; that's just terrific! Give me a few minutes and I will be right back." I started to walk towards the front door.

"Here, Sir, look. Take this lovely gilt-framed mirror for your wife's belated Christmas present, and I can't offer you another damned thing!" That was when I agreed that we had a sale. I had already figured that I had done very well.

"Very good, Sir. How are you going to pay for this?"

(This is the funny part.) "On the never-never," I replied.

"What do you mean the 'never-never'?" he asked with a puzzled look on his face.

"Well, the 'never-never,' is like—well—so much down and so much a week. Don't you have this over here?" This was British jargon for obtaining something and never being able to fully pay for it as it seemed to be a recurring situation. One thing wasn't fully paid off and you were in the process of buying something else; which meant that there was always a little bit of the first thing still to be paid, even away down the road to the "umpteenth" thing! The "never-never" described it just right.

"Oh, you mean on *credit*? We can maybe arrange to have this done."

He didn't know at that point that he was dealing with someone straight off the boat, but he wasn't long in finding out he had someone who had absolutely no history of credit. (He did give me a rye and coke in the back shop later.)

Looking back, I realized they must have been desperate and decided to take a chance on me, for the furniture was delivered to Plewes Road in a few days. All I had to do that evening was fill in a form and give them a few dollars down.

(There were still parts of that three-room suite in my daughter's basement in Winnipeg when I wrote this. I know the kitchen chairs were still in existence, having been recovered once, and the table was still in use, too. Audrey used them for her ceramics class. I think the dresser and the chest of drawers were still in her basement for storing ceramic things, so that isn't bad for well over 30 years!) There's one thing I'm very sure of—if I had been trying to buy that same furniture in the middle of, say, May or October, I would have had to pay much more than I did just after Christmas.

~ * ~

While at Hepburns, there was a little English foreman called Ron. He didn't seem to like me for some reason or another. I had overslept one morning and, to prevent myself from being late, I took my breakfast sandwich with me, got to work on time and was chewing on it while I was operating the radial drill. He rushed over to me and shouted at the top of his lungs, "What the hell do you think you are doing eating your breakfast here? In future you eat your food at home before you come to work."

I told him that I didn't think that I was doing any harm as I was working at the same time and that if I had eaten it at home I would have been late. His stupid answer (still shouting) was that I would have to be late next time! Oh well, it takes all sorts. He's probably died of a heart attack long before this with a disposition like that!

~ * ~

Despite Ron, work was fine. Things seemed to be going better. I needed a car but still didn't have enough money to buy one. I can remember walking down a street close to where we were living at Kew Gardens, when I saw an old car with a "For Sale" sign in the window. I said to Mary that it would be nice if we had "enough money to buy something like that."

Her reply was, "Dream on, Ian; we don't even hae enough money tae buy one o' the tires!" This was a car that was on a par with the *old* cars I first had when in Scotland! No, I hadn't come very far!

~ * ~

I had opened a bank account in a branch of the Royal Bank on Wilson Avenue, which was quite near to where we lived after we had moved to the basement apartment. I wanted to buy a 1959 Chevy Belair two-door sedan from an Irishman who was going back to Ireland. ('Lucky Devil!' I thought). We agreed upon a price of $360. So, I went to ask the bank for a loan.

When the loans officer asked me how long I had been working at my present job and I told him three weeks, he then told me that I had to be permanently employed for two years before they would consider me for a loan. Quite a while after this, going by the hiring and firing that was done in this country, I figured that very few people in Canada ever qualified for a loan from any branch of the Royal Bank. When I told Mr. Teoli that I couldn't get a loan to buy the car, he said that he would co-sign for me. He very kindly did this and I got the car. (Leave it to old-time immigrants to have finally learned the ropes and be willing to help the new immigrants! Bless you, Mr. Teoli!)

~ * ~

Sometime in late February I received an envelope in the mail from Wonder Bread containing a T4 slip. I didn't have any idea what it was as I had never heard of a T4. I phoned the Wonder Bread main office to find out about it and was informed that it was to send to the government when I had completed my income tax return. I then examined this form and noted the amount I was supposed to have earned.

When you don't have much, it isn't difficult to know when the figures don't jive! I said to Mary, "Hey, look here—look at the amount of money Ah'm *supposed* to hae made at the bakery job; I didna earn *that* much!"

I called them back and talked to someone in the head office for a little while and was then told to call the depot that I had worked at. He didn't give me a phone number and in those days, I, being quite new to the system, didn't ask for one from him. I then had to go to the phone book to get it. When I called, the line was busy. I waited a few minutes and called again. Again I got the busy signal. (Back in Scotland they would say that the line was "engaged.")

It was free on the next try and when I got to talk to the person in charge of that depot, I was greeted with, "Oh hiya, Ian, I'm glad you called. We've been trying to find you for ages."

This had to be a lie, because he could have got my phone number (for free at that time) from "information" and we had put in a "change of address" card to the Post Office, too (had to pay for it). Somehow I had the feeling that he was expecting the call. I talked to him about the T4 and he said that there was money waiting for me in his office.

Well, this was music to my ears! I could certainly use some of that stuff, for since we had arrived in Canada, money and I were not the closest of friends!

This happened to be a Friday (early finish at work), as I had got home at four o'clock, and knew that the depot was open until six. I told him I would be right down to collect it.

Yes sir, I got my Toronto map out and figured out where I had to go. I had my big American car and I could drive there, no trouble at all. I'd never been down that way before but I figured that all I had to do was look at my trusty map, right? Wrong!!

It wasn't too long before I was completely lost, managed to get myself "*downtown Toronto*" and during Friday rush-hour, too! Bloor Street and Yonge Street are murder even when it isn't rush-hour. I was sweating bullets and still getting used to driving on the "wrong" side of the road with this BIG American car. (Heck, it was almost as big as the dump we *lived in* when I was a boy!) There were millions of cars everywhere. And it was getting dark.

I eventually got myself off the main drag, drew to the side of the street to have a good look at my map, and even tried to flag down a passing pigeon to ask it directions (pigeons know all sorts of directions). I had noted the name of the street that I had entered, found out where I was and by this means (a couple more times) I managed to (eventually) get to the depot before closing time.

When I entered the office I was greeted as if I were a long-lost brother, getting my hand shaken by the three guys that were sitting and chatting. Then I was given a seat. The boss man then gave me a sealed envelope, explaining to me that he had to cash the cheque for me as it would have been returned to head office and then I would never have seen it again. I accepted this and left for home. By then it was totally dark. Rush hour was over, which was a blessing, but everything looked different in the dark. I was quite relieved when I started to recognize "my" part of town.

While driving home, I had a chance to reflect on the envelope I had received. I had opened it before driving off, to find out just how much richer we were. There were forty-odd dollars and some change in it. (They were all crumpled old notes, like there had been a sudden collection made.) Hey, this was a "Godsend" for us; it was close to three-quarters of a week's take home pay. An extra *forty dollars*! WOW, this was great; we could sure use it. But where did it come from?

After thinking about it for a while, I figured that there must have been some sort of scam going on. When I got home and gave the envelope to Mary. I waited to see her face light up when the money was revealed; and I was right, she was delighted.

I had already figured that the guy at the Immigration Department had been in cahoots with some guys at Wonder Bread and was probably given backhanders for getting suckers like me (just off the boat, green behind the ears and didn't know any better) to work for them. How else would he have told me that there were no openings in my category when he had the same file as the other guy which contained lots of jobs in my field? This method of recruitment most likely was not known to the "top brass" at Wonder Bread and I'm quite certain that if it *had* been made known to them, heads would have rolled.

When I had estimated the amount of money I should have earned driving the baker's van, I realized that my quitting there had not been reported, allowing "whoever'" to receive an extra pay-cheque in my name, at least until the end of the year. This, most likely, went on even after the New Year, when it would have been into another tax year. I was not aware that it was. I didn't receive any more T4s from Wonder Bread, but that doesn't mean that there weren't any. We had moved a couple of times since then. When I became aware of what had been going on, I then had another dilemma to face. Obviously, someone had forged my signature.

Should I go to the police and make them aware of what had been going on? If I did, I would have had to return the money I had just received as it obviously didn't belong to me. On the one hand, I needed the cash quite badly and wanted to keep it. On the other hand, the people who were doing this should have been stopped (number one being the guy behind the desk; second, the guy who took all the workers' cheques to the bank to get them cashed—common practice in those days—and who knows how many others were involved).

I thought about it long and hard, being careful not to tell Mary what was on my mind and get her all worried. Eventually I figured (right or wrong) that the people must have got a scare with me turning up looking for money, perhaps they decided that what they were doing was very risky and that they had narrowly escaped being caught. (Maybe they were even expecting to see the police arrive at the depot anytime!) I argued with myself that as it would *now* be stopped, there would be *nothing gained* by me reporting it, plus I would lose the money we so badly needed. (I still to this day wouldn't be caught dead buying Wonder Bread! *Wonder* why?!)

I wonder what other people would have done. I really should have reported it—but I managed to ease my conscience by telling myself that I was justified in taking the extra money. It made up a little bit for the lousy rate of pay that I got and the long hours involved in earning it!

We had managed to save the few dollars of the "wonder" money which allowed us to put a deposit down to rent a nice apartment in one of two high-rise apartment buildings that were just newly built on Jane Street, just north of Wilson; so it did serve a purpose.

We were on the third floor, with three bedrooms. There was a lovely swimming pool plus saunas in the basement, lifts (elevators) and underground parking. We could then have cablevision. The rent was $140 a month and our apartment would be ready for occupancy in July.

~ * ~

It was also during our stay at the Teoli's that Mary's brother came to see us. This would be around April or May 1966. Ken and his family were living in Toledo, Ohio, and he was on his way to Montreal to try to hire workers for the shipyard where he was a foreman. He stayed overnight with us (sleeping in a "foldout" tent trailer he had pulled behind his car, for we didn't have room for him). I can remember feeding an electrical cord through our basement window to allow him to use his inside light. (He probably had an expense account from his company and by staying and eating with us instead of a motel, would likely be able to pocket the cash.) He certainly wouldn't do it with his *own* money. You'll see why when I relate the next story!

On his return from Montreal, he again stayed with us and that was when he invited Mary and the girls to Toledo. I knew I couldn't go down as I was working and didn't have any vacation time saved up. I thought, however, it would be good for Mary and the kids to go there as it would be like a little holiday for them. We were six months in Canada and we still didn't have any money to speak of, just a few dollars in case we suddenly needed something. It was agreed that Mary and the girls would go down to Ken's in June.

Mary got all the directions regarding buses, was told to phone Ken when they got to Detroit and he would drive up there to meet them. Well, they got to Detroit. Mary phoned and waited and waited for Ken. Eventually she had to phone his home and he was still there. He told her to get the bus to Toledo and he would meet her at the bus stop. No explanation was given as to why he didn't go to pick her up. (Of course, I didn't know anything about this until it was all over.) Mary and the girls were complete strangers in a "not too good" downtown, foreign city and she had to fuss around to find out about the Toledo bus.

Then she had to phone Ken back to tell him what time her bus was leaving.

When she eventually got to his house and she phoned me to tell me that she had arrived, that was when she told me that she and the two girls had to sleep in the tent trailer in the driveway of his home. (The one he slept in when visiting us.) This was during fierce thunder and lightening storms with torrential rain. Would you believe it—this guy invites his sister and nieces to stay with him, knowing full well that there was no room in the house for them to sleep.

I was going to say that "that beats all" but it doesn't; I'm not finished. During the next day or so they *all* went to the local swimming pool. As they were lined up for the tickets, Ken said to Mary, "I'll get the tickets for all of us and you can pay me for you and the lassies later." (It was *all* of a dollar—something.) I couldn't believe anyone was *that* cheap (especially with his own sister and little nieces who had just arrived in a strange land and were desperately trying to get their lives together).

Hang on, I'm still not finished! (I know I keep saying "you won't believe this" but it is really amazing to me, even after all this time, and I *have* met some really miserable cheapskates during the years I've been over here, but this guy takes the cake!)

I was definitely wrong about no one being *that* cheap. Mary phoned to tell me that she wasn't going to stay the ten days as planned. She was returning home the following night; this was about a week into their "holiday!" After she made this decision, she asked Betty, Ken's wife (who was also from Glasgow and whom we knew very well) if she owed anything for her keep while she was there. (Mary is the type of person who expects nothing from anyone, so always offers to pay her way if necessary.) Betty replied that she had nothing to do with that and to ask Ken when he got home from work.

His reply? "Oh, just give me $65; that should cover everything." That was a week's wage for me after taxes! (Seemed like *someone* else wanted me to pay their mortgage.) Mary also told me later that *all* they got to eat were pancakes and bananas, and I believed her, because in our home we believe in always telling the truth no matter what the consequences. But it's not a matter of *how much* he charged, even if it was as little as five dollars, the thing was that he *did* charge her for her keep. I was wishing I had been there and he would have heard a thing or two!

Now, I have to emphasize that this does *not* reflect Scottish hospitality. There's no other Scot I know here in Canada (or anywhere else) who would ever think of charging invited guests for staying with them. It is unheard of, with that one exception. Whenever I invite *anyone* to stay with me, no matter where they are from, there is no thought of any reward or charges for their keep. Even if I was stone broke, I would never ask for money from anyone who was being sheltered under my roof, at my invitation. I'm sure that most people feel this way.

This brother later lived in Florida. I had to make a call to him to get a phone number of a relative in Scotland during 1996. Mary said to me after I came off the phone, "Maybe he'll invite us doon for a holiday when ye retire."

My answer to that was, "I would not be able to pay what he would charge us for food and lodging, so you can forget that. We would probably find it cheaper to go to a hotel!"

~ * ~

I may as well relate a later experience we had in Toledo. Neither Mary nor I could remember the circumstances, but for some reason we were down there a little later to see Ken about something. (I didn't have a great desire to spend *any* time with this guy so it must have been important.) It also must have been a "one-day" affair too, for there certainly wasn't room for us to stay with them and I certainly wasn't going to sleep in their trailer!

When Ken had come to visit us at May's shortly after we arrived in Canada, one of the things he bragged about was the big, l-o-o-o-ng, chesterfield he had. I was *dying* to see it as we drove to his place. I knew *everything* in the States was "bigger" but how big could a chesterfield be? Well, was I ever disappointed! When we were sitting in his house having a coffee (he only charged twenty-five cents for it. Just kidding!), I asked him to show me the long chesterfield. He replied that I was sitting on it. It was just a four-seater—a little longer than ours—but certainly not any longer than the one that May and Bill had when he was visiting and bragging.

Ken seemed to be more American than a Texan. He also spoke with an American accent. He was all of *six years there*! (Even after over forty-six years in Canada, I still speak like I'm just off the boat!) Anyway, when it came time to make our way back to Canada, Ken drew me a little map so that I wouldn't get lost. He indicated what turns I had to make and then he made a little squiggle on the paper and showed an intersection just beyond it. Well, this "squiggle" turned out to be a short, sharp hill that was in a 45-mph (72 km) zone. I went up this hill at that speed and immediately down the other side.

At the bottom was this intersection. The traffic lights were just turning red for me as I crested the short hill, and three lanes of traffic were just beginning to move on each side of that other road. I could do nothing else but put my foot down and race through that intersection at twice the posted speed limit, for if I had jammed on my brakes, I would have ended up skidding to the middle of the intersection! There was no advance warning of "traffic lights ahead" like we have here in Canada.

I was "all shook up" as the song goes, and had heard quite a few stories about the police in the States and how "trigger happy" they were. Well, as I sped through that intersection, I was expecting a bullet through the rear window—no kidding! I drove for a couple of blocks and eventually stopped somewhere in a side street to compose myself. No one was following me to arrest me. No sirens were wailing. Whew!

I was hopelessly lost—had no idea where I was. The street light system was different in some places, too. *One* light was hung exactly in the middle of the intersection and, not being aware of this, I didn't know what was happening. I was used to having traffic lights at each street corner. Anyway, I knew that I had to find the highway that Ken had told me about. I did find one and it was a *beauty*. Not only that, it was also empty of traffic. How lucky could I get?

So—I was driving along, quite composed and doing fine when suddenly I saw in my rear-view mirror a cop car heading towards me at what seemed like twice the speed of sound, with his lights flashing merrily and his siren going. I wondered where he was heading at that speed. I drew close to the right and slowed to let him pass, although there was lots of room (as we're supposed to do in Canada) and then he pulled in behind me. I got out of my car (not knowing I wasn't supposed to do this), then he got out of *his* car. He was massive—about six feet, six inches (200 cm.) tall and about half that wide. Oh, yes; he had his "shades" on, too. That was when I opened my mouth.

"Is there something wrong, Constable?"

Constable? *Well*, you should have seen the look on his face! He was probably saying to himself, "What on earth do we have here? Someone from the moon?" I guess the "Highland burr" threw him for a loop as well.

"What are you doing on this highway?"

"Oh, I hope I'm heading back to Canada. Am I going the wrong way?"

"Can you tell me how you got on here? This highway isn't opened to the public yet and all the ramps are supposed to be sealed. I had to open one up to get on here to catch you."

"Oh really? I saw a ramp leading to the highway and I assumed that it was the one I wanted as it was heading north and I just drove onto it. I didn't see any signs. I wondered why it was so empty." And I was laughing as I said that.

By this time he too had a great big grin on his face and he asked me where I came from. I told him "Toronto" and he said, "No, I mean before that. I think you're Scottish."

I told him I was, that I was just a few months over here and that this was my first visit to the States. He then told me that his ancestors were from Scotland, wished me well and told me to follow him and he would take me onto the highway that would take us to Detroit and then Canada.

When he had us safely on the correct road, he gave a little "whee" on his siren and waved goodbye. (So much for the "trigger-happy" cops I was told were in America!) I could just imagine him relating the story to his family during dinner, saying something like, "Wait 'til I tell you about the little Scotsman I had to chase down today . . ." Or his fellow cops at the Precinct—they would have got a hoot out of that one!!!

* ~ * ~ *

Toronto, our rental house on Stuart Smith Drive. Shirley helps to paint. Audrey "supervises" from the front door.

CHAPTER THIRTEEN
We Move to an Apartment and Then to a Rental House

We took possession of our brand-new apartment August 1, 1966 and thought the underground parking was absolutely wonderful, especially in the winter when it was icy or snowing. It was even better when there was freezing rain, as the car windscreen (windshield) was always clean. We all liked the big swimming pool and the saunas down in the basement.

Only a couple *comical* situations arose there, mostly regarding my car. I was driving home after an evening of shopping with Mary and the girls. It was dark, and as I was heading north on Jane Street from Wilson Avenue, my passenger-side headlight seemed to go out. I told Mary that I thought I would have to get a new lamp the next day. When I drove into the underground parking, the light shining at the front of the car looked "funny." When I reached my parking spot, I left my headlight switch on and got out of the car. There was nothing wrong with the lamp itself. It was lit, shining *at* and just a few inches from the ground and hanging by its wires. What was wrong was that there was no metal left to hold the assembly; it had all rotted away.

This was a seven-year-old car. It was *then* I realized why the used cars in Toronto were so cheap! The metal would soon be eaten away as a result of the road equipment teams "salting" the icy streets in winter to make them safer to drive on.

The next day I spent making brackets and fixing some supports for the headlight again. After applying some fibreglass to fill the empty space where the original metal used to be, then smoothing it with sandpaper and painting the spot, it was good enough for me.

I discovered a lot more rot just a few weeks later. I was in a small strip mall and it was just about dark outside. I noted where I had entered, and at the other end was a wide space, so I assumed that was the exit. I drove out that way after my purchase was made and found out it wasn't the exit at all! It was just a wide space where there were no cars parked. When my front wheels dropped off the curb, I immediately heard a loud "clunking" sound. Then my back wheels came off the curb and there was another loud clunking sound. I stopped the car, got out, surveyed the situation, opened the trunk and proceeded to collect the two rocker panels that had just fallen off, one from each side of the car just below the doors.

A few weeks later Mary and I were at a dance when the emcee announced a license plate number and asked the owner to please go outside as the car was on fire. Well, the number sounded too close for comfort! I went outside, and sure as fate, it *was* my car. The wiring had shorted and *I should have let it burn*; but "silly me" had to rush to the trunk, get a wrench and remove the battery connection. I did fix it the next day but I knew the writing was on the wall for *that* car. It was seven years old and I had only paid $360 for it. I had finally realized why it had been so cheap!

~ * ~

We used the swimming pool a lot. I guess I was starting to go bald by then as my girls said they could always spot me in the pool by looking for the shiny spot at the back of my head! I've never gone completely bald—but you'd never know it from a distance if you see me from the front. My thicker hair is mostly at the back.

~ * ~

Things weren't going too great, however. We had been in Canada for just over a year and I had just started my *fourth job*. Now, it was just around this time that Canada adopted its official national anthem (on April 12, 1967).

"*Oh, Canada, our home and native land. True patriot love in all thy sons command. With glowing hearts we see thee rise, The true North strong and free. From far and wide, Oh, Canada, We stand on guard for thee. God keep our land glorious and free. Oh, Canada, we stand on guard for thee. Oh, Canada, we stand on guard for thee.*"[10]

I didn't feel at all patriotic at the time and one thing I'll be forever grateful for is that the song wasn't adopted on *April 1st*. Man, if it had become official on April Fools' Day, I would have been sure that it was being directed straight towards me!

~ * ~

While we were in that apartment, Mary's sister Nettie and her future husband, Gillies, came to Canada for a visit. (Nettie had been widowed some years before that.) Gillies had a sister who lived in a small town called Mitchell just west of Stratford, Ontario. We found him to be a very nice person. Although he was a good deal older than Nettie, I was quite sure that she couldn't have made a better choice. While we were walking in a park one day, Nettie got me to one side and "sort-of" asked for my blessing, saying that Gillies had asked her to marry him. This was *before* she told her sister, my wife. I told her that I was not the least bit concerned about the difference in their ages; what really mattered was how happy they were together. Also, before they left to go back to Scotland, Gillies spoke to me regarding Nettie, to reassure me that he would look after

---

10 "Oh, Canada." English lyrics quoted herein are by Robert Stanley Weir (1856-1926). Original French lyrics by Sir Adolphe-Basile Routheir (1839-1920). Music by Calixa Lavallée (1842-1891).

her. I told him the same thing that I told Nettie. (Nettie and Gillies *did* get married in 1967 and came out to visit us again in 1969).

During the first visit we decided that we would go to the States for a short visit as Nettie had gone to the trouble of getting a travel visa for the USA. We (Mary, Audrey, Shirley, Nettie and I) all went for a day trip to Detroit (230 miles or 370 km) to let Nettie make her first visit to the great United States of America (a must for "talking about" with her friends after her return to Scotland and also to do some shopping at the same time).

I drove around and around, looking for a place to park and eventually settled for leaving the car in the middle of a deserted bit of waste ground not too far from downtown. I had no idea of the seriousness of what I was doing, and didn't give it any thought. There were quite a few black people sitting against the walls of the buildings that were at the edge of this bit of waste land.

They were looking at us but I didn't pay very much attention to them as, to me, they were just people sitting there chatting to each other. I, being me, most likely gave them a wave and said "hi guys" or something like that, for it's well known that I'll talk anyone, "even to the Devil himself," my wife says. We went off to the stores, did our shopping, returned to the car and drove back to Canada.

The following day we took Nettie down to Niagara Falls. There, we rented waterproof hooded capes, went under all that water and had a genuinely nice time. Before we headed for Buffalo, we stopped at the side of a nice grassy area at the side of the road, opened the trunk to set up for a picnic and the next thing we knew the grass was alive with snakes! There were hundreds of them! Everyone made a mad dash to get inside the car! I told them that they were only harmless grass snakes, but that made no difference to the females. They said we should find some other place to have our picnic. I didn't see the reasoning for this as all the snakes had left us by that time.

We did quite a bit of shopping after we got to Buffalo and it was getting dark, so it must have been a Friday when stores were open later. We talked about trying to find a motel to stay the night and ended up with all of us sleeping in the car overnight with more shopping to do the next day. The first item of business in the morning was to find a service station, get to a washroom and also to get washed and shaved. Then it was breakfast and some more shopping for Nettie to get excited about.

I had been on vacation and when I returned to work I was relating this "adventure" to some of my workmates. They remarked that we were very lucky that we still had a car when we got back to it at Detroit and also lucky that we were still alive after being all night in the car in Buffalo, as there was quite a lot of rioting going on between blacks and whites in both of those cities at that time. The black population was protesting about discrimination, and apparently our car could have very well been wrecked, never mind what might have happened to us. After I thought about this, I figured that our car wasn't touched because it was an old rusty thing with *Canadian license plates*; therefore it wasn't belonging to "the enemy," so to speak. The American blacks didn't seem to have any arguments with Canadians.

~ * ~

I worked at Hepburn's for a few months and then was laid off. (There weren't nearly as many layoffs in Britain as they had in Canada.) The next job I had was in a custom sheet metal shop in Downsview called Embury's, making stainless steel doors and frames, shop fronts and commercial kitchen fittings. Lloyd, the boss, was one smart cookie. He, apparently, had invented a machine that fed out anodized aluminum from a selected coil, with a measuring meter and guillotine incorporated. A little after I started work there he completed a much more elaborate one that held twice as many coils. Not too long after this there were quite a lot of Japanese people in the shop, going around the machine, taking pictures, measuring, etc. He then sold the Asian rights for that machine.

He designed and built a 20-foot (6-meter) 200-ton press brake. It wasn't that marvelous that he did this, the important point was that its *cam* system was much better than others. This was a machine for folding long sheets of metal.

Also, during this time he was working on a machine that seamlessly welded and then ground smooth 2-inch wide (5 cm.) stainless steel banding, rolling it into a tube which reduced in diameter as the tube grew in length. It ultimately resulted in a constantly tapering flag pole (something which hadn't been in existence before). He had just perfected it shortly before I left there to work at Douglas Aircraft. The thing I liked most about Lloyd was the fact that he was the first to admit that all he had was a grade eight education.

(During this time we had moved to an apartment on Jane Street, Toronto.) A new man (Don) had just started work. We got along well, always managing to have a laugh together. One day he asked me if I was interested in renting a little house very cheap. I asked him "how cheap?" and he replied "around $60 a month." *That* was cheap; not nearly as cheap as our nice house in Scotland but compared to Ontario it was and I asked him to tell me more. It seemed that there were two small houses (520 square feet each but with full basements) right next door to his in-laws. The houses belonged to a steel company, the rear of which backed onto the opposite side of the cul-de-sac (dead end street) where the houses were.

The one next door to his wife's folks was empty and it was a little too close to them for Don's comfort, so he thought he would give me the chance of getting it. I spoke to Mary when I got home and she thought it might be good. I couldn't wait, and as soon as dinner (or I guess I should say "supper") was over, we all went down to see it. It really *did* need work, lots of it, and the back door was open wide to the world. This was good and it was also bad. It meant that we could get inside to see what was needed to fix it up but it also meant that others could get in and do a lot of damage to the property. I couldn't allow that; so I secured the door to prevent others from gaining entry. The house also had a front driveway.

As Don had given me all the information I needed, I got away early from work the next day to contact the person responsible. I found out Don was a little out in his estimate of how much the rent would be.

When the man in charge said that he wanted $100 a month rent, I told him there was no way he would ever get that, recommending that he go and have a look at the property and then we could talk some more about rental charges. Apparently he hadn't seen the house (inside or outside) for years, so he was probably "seeing" it as it had been long ago. The next day when I again left work early to talk to him, he admitted that it was in a terrible condition. I made him an offer. I told him that I would get the house back to how it should look, supplying my labour free. For this he had to agree to let me have the place for 80 dollars a month. It was agreed that I would purchase whatever was needed—wallpaper, paint for outside and inside, tiles for the bathroom floor and, in general, any wood and whatever else I needed. Also, the first two months would be rent free as we would have to remain in the apartment while I was upgrading the house to make it fit for occupancy. I also required immediate reimbursement for all costs, including purchase of a lawnmower, as soon as possible after presenting the bill to his office, since I didn't have that kind of money to throw around. Getting reimbursed immediately allowed me to purchase more material. He agreed on all of those terms and, after a lot of work, we moved into the small house at 42 Stuart Smith Drive (now gone), just east of Jane Street.

~ * ~

We found about a million empty wine bottles in the basement of the new house—well, maybe 300, not much less! After we had taken over the house, I had to put those bottles out for the garbage men to collect. I couldn't do it all at once, so it took weeks and weeks, a couple of dozen at a time until they were all gone. I wonder what the garbage men were thinking. (Now that I make my own wine it would have been nice to have some of them back!)

~ * ~

At that time, we had close friends, John and Rena Marline (from Glasgow). John had told me quite often how crazy I was to rent such a small house. My reply to him was that it was the next best thing to having my own place. The girls shared a bedroom and there was a full basement for storage. I could drive a nail into a wall and not have to ask anyone for permission and other things like that. Essentially, the house looked worse than it really was—I was surprised at just how little the inside needed initially to get it into reasonable shape.

Mary, the girls and I were able to move in within a month, so this gave us another month's free rent. I got more time for wallpapering and the like. When I had the inside all done, I started work on the outside. I had met our next door neighbour, Rusty, quite a few times (he was in the other small rental house belonging to the steel company, and the last one on the street). Russ turned out to be an alcoholic who lived alone. (Maybe the 300 wine bottles had come from him!) When I had got around to getting the outside all painted, this guy wouldn't talk to me. I asked him what was up and he told me that before I had moved into the house at #42, it was the worst looking house on the street, now *he* had the worst looking house! It was true, too!

However, our living there was not what could be called a "long term" situation. The two wee houses were slated for demolition whenever the steel company was able to buy the house on the corner where Don's in-laws lived. They were not willing to sell until they got the price they wanted. The steel company planned to eventually build a high-rise apartment building and Don's in-laws on the corner were holding the project up. It appeared that exactly where their house was sitting, was where the entrance to the underground parking was to be. So, as soon as they sold to the steel company, those three houses would be torn down.

~ * ~

It was while we were living there that I got a call at work from Shirley, our youngest daughter.

"Dad, Audrey has cut herself with a knife and she's lying on the floor not moving!" Then she hung up. I don't know why I didn't get a speeding ticket! I think I was home in about half the time that it usually took me from work. It turned out that Audrey had been cutting an Easter egg when the knife slipped and cut her thumb. She has no head for the sight of blood and had just passed out; but as I was driving home I had all sorts of things going through my head. She was fine when I got to the house. One thing we knew about Audrey for a long time was that she would never ever be employed as a nurse.

~ * ~

Dennis Hopewell was the man initially responsible for getting me into the Royal Canadian Legion in Woodbridge, Ontario. Dennis' wife, Jessie, and Mary used to work together at one time and we had become good friends. They were English, and a nicer couple you couldn't hope to meet. Jessie and Dennis already had a house in Malton.

There was an excellent bunch of people there and I should have remained with them instead of transferring later to the "Rangers Branch", close to where the little house was, somewhere around Keele and Eglington.

Our branch at Woodbridge had an old piano that they wanted rid of. It was to be given away free, the only clause being that whoever wanted it should remove it by a certain date. Its condition wasn't too bad. It sounded all right, the pedals worked (not that I ever used *them*), and it didn't look that bad either. I rented a U-Haul trailer, picked up the piano and took it home (at that time to #42 Stuart Smith Drive). It was to go down in the basement, and fortunately the stairs were right in line with the back door.

Another thing in my favour was that there was a fairly large tree also in line with the back door and this allowed me to wrap a rope around the piano (which was balanced at the top of the stairs) and twice around the tree. I controlled the rope, easing off a little at a time, and a friend tilted the piano to get it heading downstairs. A piece of cake—it went down so easy you would have thought that the two of us were experts at the game. Mary's cousin, May, her husband, Bill, and the two of us had a few good sing-songs around it in the years to come. By the way, the piano was taken out using similar tactics when we moved to our new house.

~ * ~

About a year after we moved into the little house, I got what I thought was a great idea—transferring to the Rangers Branch to save the long drive to Woodbridge. There is something I should explain here to shed some light on what happened after the transfer.

Rangers is a Scottish (Glasgow) soccer team. Very strong feelings have traditionally existed between them and the Celtic (pronounced "seltic" on this occasion) soccer team, also in Glasgow. Rangers' fans are strongly Protestant and Celtic fans are strongly Catholic. So much so that there has been quite a lot of blood spilled (literally) over there during games (sometimes before, and sometimes after, too). This violence is done *by the fans*, and has been going on for years. It was a strong boast that a Catholic would never play on the Rangers team, whereas Celtics had no such claims (a bit more sensible).

I put in a transfer to the Rangers branch and never once did I associate the name with the soccer team in Glasgow. Even if I had, I certainly wouldn't have linked it to any bigotry. I couldn't really be called a fan of either team.

Three months had gone by and my transfer still hadn't gone through. After I had attended the third monthly meeting at the new branch (while waiting for word that I had been officially transferred), I rose to my feet after all other business had been concluded, and asked how long it normally took for a transfer to take place. There were a few murmurings around the floor such as, "Who does this guy think he is?", "What is he talking about?" and other remarks. When I made it clear who I was and that I had been to two other meetings and no one had spoken to me during those meetings, that I had applied for a transfer from Woodbridge and that I thought it was a bit ridiculous that it should take so long, I was assured that it would be seen to. (This is not the end of the story!)

Shortly before Christmas, a Children's Christmas Party was being held at the branch. I put Audrey's and Shirley's names along with their ages on the necessary form. When it came time for the kiddies' party, my two girls were almost left out. They were given their "presents" *last* and each amounted to about a ten-cent piece of junk—I'd have thrown better into the garbage! That was when I overheard a couple of "old time" members talking about me. (I think I was meant to hear.)

"Yes, his name is *Moran*; that tells you right away that he's a 'Mick'. He'll get what he deserves." They were *driving me away* from "their" branch; for they didn't want any Catholics there—just like the soccer team in Scotland! I couldn't believe that such stupidity existed in Canada. Unfortunately it did and it was taken there by *my own* people—I was disgusted!

A "Mick" is another name for an Irish Catholic. (A "Mac" is Scottish and usually Protestant.) They thought that *I* was Catholic and were branding me accordingly and also punishing my two little girls at the same time. That was it for me! I searched for and found the president of that branch and told him about the stupid people he had in his branch, probably him amongst them, and exactly what I thought of their treatment of people according to their names. I informed him that they didn't even have my name *correct* for starters, but were still willing to persecute my two little girls and me; concluding with the fact that I was *Protestant* and he and his kind were not making me very proud of it! I told him that sensible people left that sort of bigotry back home when they left there. Then I told him to stick his transfer and his "gifts," which I threw at his feet, and that I would think twice if I were ever to join another Royal Canadian Legion!

Makes one wonder, doesn't it? I related the story to the head office and got an apology, but not from the Ranger's Branch.

~ * ~

We had continued to keep in touch with Jack and Jean Pepper. Jack was working as a shop foreman in some factory that made "U" bolts for car exhaust systems, among other things. He gave me a phone call one evening and asked me if I would be interested in taking a job at the place where he worked. We talked some and I said that I would think about it. He said that they badly needed someone who could repair the machines in his department. I went to his work place one day and talked to one of the bosses who assured me that there were great things ahead for me at their company. I quit Embury's and took the job.

I soon found that there was not a dull moment at that place. The machines were in such bad condition that they were continuously breaking down.

I was only there five days when Jack told me that we were called to a meeting.

"When we go in here, Ian, don't be afraid to speak your mind. If you have anything to say, then say it."

We were all seated and the manager told us that the reason we were called to the meeting was because the company was very far behind in its orders and that we should discuss any possible way of trying to improve our output.

"For instance," he said, "we are 27 million 2-inch U-bolts in arrears, and that's not counting the other sizes." (2-inch was the most common size).

The company supplied all the big auto manufacturers (in the States, too), as well as Canadian Tire and other large outlets.

"Well, Jack, seeing as you're the man in charge of that department, what do you think we have to do to improve things?" Jack just shrugged his shoulders and said nothing.

There was a young man, Ralph, who worked with us and was usually quite resourceful. The boss asked him if there was anything that he might think would help. The young fellow replied that he had no idea.

"I don't think it is quite right to ask you, Ian, if you have anything to say. You have just been with us for a few days; but if you have any suggestions as to what we could do, I would certainly appreciate hearing from you."

"I know exactly what your problem is," I said. "The question is, will you believe me and do what's needed to improve production?" You would think that I had just given him a million dollars.

"Say what's on your mind, Ian, and then we can all talk about it while we're sitting here."

"The problem is the condition of your machines. They are in such bad shape that they are continuously breaking down. I've been here five days and haven't had time to enjoy even a cup of coffee at break times. Even during lunch, as the girls on production are on split times, we don't get a full half-hour lunch break. One of us will be working on a machine that's just broken down and a different girl is calling for help for another machine. It's the same with all the machines all the time. All we get time for is running repairs and that's no good."

"That's the best we can do, Ian. We have to get what we can out of the machines. I know they are not in good condition but what else *can* we do?"

"Sacrifice one machine at a time. Shut one down completely; forget about it as far as production is concerned; get the toolmakers to rebuild it, making new parts as necessary. When that one has been rebuilt and in full production, then shut another one down and do the same; then the next and the next and so on until all the machines are in good condition. Then you will have full production. You must also convert *all* of the machines to make 2-inch U-bolts, if there is such a massive deficit in that one size. That's the only way it will be reduced. That's what I think; now it's up to you."

We all discussed what I had said and decided unanimously that this was what was needed. The manager said that it would be talked about at higher levels and that he would let us know what decision had been made. No doubt, I thought later, that this "idea" would be *his* idea, with no credit going my way; and I was correct. If he was "production manager," then he should have known how to fix the problem. That is what those guys got the big bucks for, not for picking other peoples' brains. After we came out of the office, I asked Jack why *he* hadn't said anything after telling *me* to speak up. He replied that he really didn't have any idea what was needed. Maybe he *did* know but preferred to keep his mouth shut. I know it took me some time to learn that lesson.

The next day Jack came up to me just after lunch break and asked me if he could borrow my car. It seemed that his car had a flat tire that morning and he had taken the bus to work.

"Sure you can, Jack, where are you heading for during working hours?"

"I've just been fired and I want to go to Unemployment Insurance (now called Employment Insurance) to see what jobs are on the board."

He was welcome to borrow my car. I told him that I was sorry that this had happened and that I hoped it wasn't because of what I had said in the office the day earlier. He replied that he felt that this was in the works for some time and that he was glad it was over. This was when I questioned his wisdom again (thinking back to the time they bought wine for us on board ship when they could least afford it). He had persuaded *me* to leave my previous employment to work in a place he *didn't feel good about*. Then they seemed to just fire him for no apparent reason. I thought, 'demote him, if needed, but they shouldn't fire him.'

The following day I was asked to report to the manager's office and found out they were offering *me* the foreman's job. At first I refused it, feeling that it was because of me that Jack was out the door. The manager assured me that it wouldn't have mattered whether I had come to work for them or not; Jack just wasn't suited to the job. I took a couple of days to decide and finally accepted. Despite this, Jack and I remained friends and continued to visit each other.

I worked all sorts of hours there but still didn't get any more money for my labours except for the overtime that I earned. I had started work there at a little less than I had been earning at Embury's, but was *assured* that the money would be there. I was working until nine o'clock every evening, trying to remodel the flow system of the blanks, designing and welding conveyor rollers to assist the young women to save them from lifting the heavy boxes. I built a conveyor system so that the containers rolled on the conveyor from the time that they entered the department until they went out. No more heavy lifting. What did I get for it from the women? They were worse than kids! They would fall out with me for no reason and when I asked them to work overtime, the answer would be, "We'll work if Mrs. Ruben asks us."

My answer was, "You either work for me or you don't work at all." That sorted them out as they were all needing extra money.

I was so preoccupied with the problem of increasing production that Mary and I were walking through Yorkdale Shopping Centre one Saturday afternoon when suddenly she hit me on the shoulder.

"Hey, what did ye do that for?" I asked, jokingly.

"Ah've been talking tae ye for the last five minutes and ye havna heard a word Ah've said." And she was quite right; I was *too* engrossed in my thoughts. There wasn't any more money for my hard work, not counting the little inventions I had installed to increase production. Whenever I asked about getting an increase in pay the answer always was, "Just wait a little, Ian; it will come. Right now we are still trying to make money."

Well, I was sick of hearing this and sick of trying to make things better for them and not getting any recognition for it. In a few short months I had reduced their deficit by eight million 2-inch U-bolts.

~ * ~

One day Audrey fell and broke her wrist and I had to take her to the hospital. On the way back home, I had to pass my old work place (Embury's) and thought I would pop in to see some of the guys. As I walked in the door, the first person I met was Dorothy from the office.

"Oh hi, Ian; have you come to apply for the job in the paper?"

"Not actually," I replied, "but if you have a job vacancy I'll apply for it, seeing as I'm here."

Well, as I had previously quit there, I had to see the boss man himself about the vacancy. I had to wait about half an hour and when Lloyd came in and was made aware that I wanted my old job again, he asked me what I had been earning before I had quit. I told him and he started me back at twenty-five cents an hour more.

I'm afraid that didn't go down too well with the people at the other place when I told them that I was quitting and going back to my old employer. One of the owners there— Mrs.

Ruben—had always been very friendly towards me (like I was a long-lost son). After I broke the news, she wouldn't speak to me. If you're "doing" for some people, you are "A-okay," but as soon as you show that you have a life of your own to lead, then they don't like you as much. In fact, they don't like you *at all*. Those people aren't worth knowing. My advice is: watch out for them; they'll use you.

~ * ~

I worked at Embury's for awhile and then I got a job at Douglas Aircraft in Malton, at the northernmost end of Pearson International Airport.

When I told the Embury's foreman, Bob Waters (from Liverpool) that I was quitting, the boss came to me and offered me another "two bits" an hour if I would stay. I had no idea what "two bits" meant until he told me that it was 25 cents. In those days, that was quite a fair offer.

The job I had secured at Douglas Aircraft (the Canadian division of McDonnell-Douglas in the United States) was making parts for the wings of DC-9 and DC-10 aircraft. I knew that the aircraft industry was infamous for laying off workers as soon as there was a slowdown. When I was being interviewed for the job, I asked about layoffs. The answer I got from the top guy in Personnel was, "there will not be a lay-off for at least ten years." I accepted the position and left Embury's.

I have come to accept that there are very few people in industry that will tell the truth. I don't know if it's a reflection of what goes on down in the U.S.A. but it seemed to me that it was getting more and more difficult to believe anything that was told to me by a prospective employer. The reason I'm saying this is because, instead of the ten years' work that was promised, I was laid off after ten months. (Mind you, it was also getting harder for employers to believe all that was on a person's résumé, too. So who can blame whom?)

~ * ~

Around this time, Nettie and Gillies came out to Canada for another holiday. They were now married and it must have been during July 1969 because both Gillies and I were glued to the TV watching Neil Armstrong's moon landing (July 21st). I certainly remembered the time in Scotland when I talked with my workmates about moon landings and got thoroughly laughed at!)

Aunt Liz and Uncle David also came out for a vacation at about the same time, for the visits overlapped. (Later, neither Mary nor I could remember where we all slept!) But this part of the story isn't about them; no, it's about Mary's brother Ken from Toledo. He drove up to our place to pick up Aunt Liz and Uncle David to take them down to his place for a little while. He arrived on a Sunday afternoon, around 2:30. Mary asked him if he was hungry and, since he was, offered him a sandwich. A few hours later it was dinner time and we all took seats around the table. Ken remarked, "I thought that was my dinner I had earlier."

Then I retorted just about as quickly as I could, "No Ken, we poor destitute souls up here prefer proper meals, like roast beef, potatoes, veggies—things like that!" Mary gave me a row later for saying that, but the opportunity was too good to miss! (I wonder if Ken and his wife fed their visitors nothing but pancakes and bananas during their stay, too!)

* ~ * ~ *

Our first purchased house (brand new), Hillsburg, Ontario, 1970. "We've come a long way!"

CHAPTER FOURTEEN
At Last We've Made It!

While I was working at Douglas Aircraft in Malton, (northwest metropolitan Toronto), I had been getting quite a bit of overtime and had been putting as much money away as possible, but not as much as I would have liked since I had had to buy another car. Anyway, I was looking at the classified ads in the newspaper one Saturday morning, and happened to see a house for sale for $6,000. I didn't think it was possible to buy a house for that little. It was in an area called New Toronto. (Sounds nice, doesn't it? It wasn't!)

"Here, look Mary," I said. "We can afford a house this price, canna we? It probably needs a whole lot o' fixin'; but that'll be no bother." She agreed and I phoned the realtor. I got through to the office and the girl took my phone number, saying the realtor would call me. I waited for ages and finally I got a call from him. We made an appointment to meet at the house right away. I got the address and instructions on how to get there. Then Mary and I drove there as fast as we could. As we headed to our first "house-buying" experience, the realtor's last words rang loudly in my ear.

"You should get here as soon as possible. This is an excellent price and won't last long on the market."

The realtor was waiting for me. "I think you are just too late, Ian. Someone was looking at the house just half an hour ago and has submitted an offer. We'll have to wait and see what happens—whether the owner accepts it."

It had taken us about ten minutes to get to him. He was the agent for the property and he obviously knew that this was going on when he called me. He was probably "stacking" prospective buyers, most likely the reason he was so long in returning my call. The reason maybe, was that he didn't expect an offer so quickly. Either that or he used his call to *me* to motivate the people who were inside viewing the house, as he probably had to use the owner's phone (no cell phones in those days) and he probably made sure that the "lookers" were listening when he called me. Anyway, he went back into the house to the owner and in a little while he came out to tell me that the offer had been accepted.

It wasn't a great house, but for the price it seemed to be a real bargain. (On the other hand, maybe it was too low and would have ended up being more of a headache.) It was a tall, slim, semidetached old house very much like the place we first lived in at Kew Gardens; but it would have been a start for us. It is most likely that this property would be worth today more than $200,000! If I had that amount of equity today, my wife and I would be able to have a trip around the world!

What this did, though, was to start me off on the road to thinking, 'Well, maybe we *can* get a house!" We didn't have any more money, but the seed was sown. I then spent just about every Saturday morning, when I wasn't working overtime, going through the real estate section of the *Toronto Star*. I always had read *all* of it by at least Saturday evening.

I wasn't what you might call "particular" as to where a house was located, except that I had to make sure that Mary could get a bus to her place of work (assembly line at a factory). I covered all of metropolitan Toronto as I knew many people from far and wide commuted to work. The paper was laid out in specific areas as most people who were looking to buy wanted to concentrate on their preferred part of the city instead of having to wade through all of the real estate section, which was immense. Me? I was just looking for a bargain—anywhere.

I phoned about a three-bedroom, two-storey house in Oshawa (where the General Motors plant is located). The real estate lady told me that it was an older home badly in need of repair. The asking price was $19,900. It had been reduced to that low price to get it sold fast. "With hard work and some money," she said, "it could be worth a lot more." She told me that the previous occupants had *removed the living room carpet* and that was about the biggest expense we would have. We then agreed to meet the next Sunday afternoon.

I had previously estimated that perhaps we could have the $2,000 (10%) down-payment if we had a few months to save for it. I was also told that it generally was about three months after an offer was accepted that occupancy was made. This would give us the time that we needed. (I still had a lot to learn about real estate.)

We headed there on the Sunday afternoon, followed her directions and soon found the house. We were a little early as I had never been to Oshawa and wanted to be there on time in case I first got lost. As I didn't, we had arrived early. That gave us a chance to look around the place, which was vacant, before the realtor showed up. It really needed a lot of work. The front and back yards were full of weeds and had been badly neglected, needing some topsoil and re-seeding to get them up to par. There was a real need for paint, but the house itself looked as if it could eventually be made to look quite nice.

She arrived on time, introduced herself, gave me her card and unlocked the door for us. Now—there are things in life that we know are definitely sent to try us. And there was something there that did it for my wife, though not in a favourable fashion. Someone had left a "deposit" right in the middle of the living room floor, instead of using the toilet! How this had been possible I don't know; the doors, front and rear, had been locked and there were no windows broken. Of course, that was the decider for Mary—it put her right off! I tried to explain to her that this was really of no consequence; as there was no carpeting on the floor, there would be no residue to worry about since it could be well disinfected.

The real estate agent had been talking to me about real estate as we were being shown around the house, asking the usual questions, I suppose. I told her that we weren't very long in the country and in the process of trying to gather some money together, but we really didn't know what was required to buy a house. She told me that she would be able to get me the house for *no money down*. All I had to do was apply for a second mortgage (she had the source), using that for the down payment. What money we had saved could then be used for expenses. I assumed that she meant carpeting, paint, etc. I had also made sure that Mary would have no problem getting to her work in downtown Toronto as there were regular buses.

Well, try as I might, there was no way that I could get Mary to let me put in an offer for the house. As it was to be *our* house, it needed the blessing of both of us. There was no way that I would ever go ahead with something like that without her okay. Why the hell did that someone have to shit on the floor? That person had cost me a lot of money eventually. On the other hand, why hadn't we been late instead of early? This would have allowed the realtor to remove the deposit—as I'm sure she would have done—and Mary would never have known the difference.

So it was back to pouring over the "*Star*" again. Maybe I would find another good deal; but I doubted very much I'd find another deal like that. It would have been a great "stepping stone" for our next "better" house. It too, today, would have been worth a small fortune.

~ * ~

There was a news article in the real estate section a few Saturdays later about a small community to the northwest of Toronto. A few people had recently moved into brand-new houses in a new subdivision. They were busy touting the good things about the little village of Hillsburgh. It told of houses being built for people who didn't have enough money to buy a new house using the conventional method of 10% or more down. These houses were being sold for a few hundred down and a few hundred later on occupancy. 'Hey,' I thought, 'I've got to see more about this!'

The realtor was very eager to let me ask all the questions I wanted to. "Yes, it isn't too far from Toronto and there is every possibility that your wife would be able to get a ride to her place of work with someone who was already living there. For instance, there's a man who has just bought a house in Hillsburgh who is some sort of manager with a printing company in downtown Toronto. His name is Peter Jeans and no doubt he will be able to give your wife a ride."

He told me that he would be just too pleased to pick us up the next day (Sunday) and drive us up there to let us see what the subdivision looked like. I said that would be all right and we then made arrangements.

One o'clock the next day he was at 42 Stuart Smith Drive. He took us past Douglas Aircraft and noted the time on his watch. It was maybe a few weeks later that I figured that this guy must have been breaking the sound barrier. Mary sat between him and me in the front seat, and probably I was too busy yakking to him to pay much attention to his speedometer. He had a big comfy brand-new Buick that made us feel as if we were riding on a cloud. It certainly was a half hour to Hillsburgh at the speed he had been driving. I got to know later that there was no way I'd ever be able to do it in anywhere near that time and still prevent myself from going to hospital or worse. (Forty-five minutes was the best I could do to Malton, and that was doing a good bit more than the speed limit.)

Also on the way, I asked him if the place was in the "snow belt" and his reply was, "Just on the edge, Ian; you'll be all right there."

(The snow-belt is a stretch of land just south of three of the Great Lakes - Huron, Michigan and Superior. It gets very heavy snow falls compared to the surrounding area. He lied through his teeth. It turned out to be right in the *middle* of the Snow Belt. (This country does breed some unscrupulous people who know how to stretch the truth or tell out-and-out lies—immigration officials, real estate agents, co-workers, you name it. Alas, I was destined to meet a lot more!)

~ * ~

Anyway, I'm getting ahead of myself. We liked what we saw at Hillsburgh (ends with the sound, "urg," not "urra" as we in Scotland would pronounce "Edinburgh").

The realtor must have had a deal of some kind going with an individual already living in the subdivision, as he went to a particular door and asked if it would be all right if he showed us what the houses were like inside. We were let into this home and found it very nice. It was just what Mary would have given her eye teeth for—a nice brand-new house. Then he showed us some plans that he had *inside* the show home, which was already sold. Funny enough, it was the same model as "ours." So why did we have to go to someone else's home to see the interior, unless we were needing some "encouragement" from the owners there? (Which we *did* get, and plenty, too.)

Then the realtor explained to us that he would be able to put us into a nice side-split, three-bedroom house the same as the show home (1050 square feet bungalow) for $500 down right away, and $650 later to make a total down payment of $1150, which was just a little more than five percent. The lot was 60 X 140 feet. There was a communally-owned water filtration system to ensure a fresh supply of water so that we didn't have to have our own well; but it did have a septic system. This seemed excellent. He showed us one of the few empty lots that were left, pointing out other houses that were in the process of being built or already occupied by workers from Douglas

Aircraft. He left Mary and me to talk things over. We both thought it was a wonderful opportunity to own a nice home and told him so. He then took us back to his office in Toronto, where he filled out a purchase agreement and we gave him a cheque for $500.

~ * ~

We were still living in the small house on Stuart Smith Drive. Our Glasgow friends, John and Rena Marline, had visited us quite a few times. Each time they came (and I waited for the remark coming), he always told me that I was crazy to live in the small house, remarking that we didn't have enough room. I told him that it was really big enough for all intents and purposes; that I was renting it but that I would never dream of *buying* a house so small.

The funny thing about our association with both John and Rena *and* Jack and Jean Pepper, was that the guys were earning a lot more money than I was and were getting a lot more overtime than I did. John was in the Carpenter's Union and making a lot higher hourly rate, almost *double* my rate of pay. Obviously, I was the least likely to be a homeowner. However, out of the three of us, I was the only one attempting to buy a house.

The week (yes, it took a week at that time) of waiting that we had to suffer to find out whether we qualified for a mortgage was all worth it when we got word that our application had been accepted. This was in early April 1970 and our possession date was July 1st. That gave us almost three months to get the rest of the money together, plus an area carpet for the living room floor. The houses had lovely hardwood floors throughout. The price of the house was $21,700.

We drove there every Saturday or Sunday to see how our house was progressing. At first there was nothing except four stakes with little flags on them; then came the beginning of a hole in the ground; next time it was a great big hole. Eventually over the weeks it looked as if a house was being built there. I photographed the progress.

The whole world seemed just fine; it seemed everything was working out well for us by then. We had a new house to look forward to and that kept us in an elated mood for some time.

~ * ~

It was only a few weeks after we had been told that we had qualified for the mortgage that I went in to work a night shift one night (starting at 10 p.m.) and was told that I had to report to the Personnel Office.

'What the dickens does someone from Personnel want to talk to me for at this time of night?' I thought as I headed through the maze of buildings.

When I got there I found that I wasn't the only one who had been summoned. There were umpteen of us waiting for the office to open its doors to us and many more appeared the closer it got to ten o'clock. As the door was unlocked, we filed in, were taken to a large meeting room and told to sit down. Then the personnel manager gave us the bad news. We were all to be laid off, effective *that night*, end of shift! There were more than one thousand people from different sections of the plant being put out of work. The reason, we were told, was that the first Douglas DC-10 aircraft was about to go through its test flights and there would be no work for us until they found out, I guess, if the aircraft was any good. So much for the ten years' work that I was told I would get; all I got was *ten months*!

I still had six weeks to go before moving into our new house. What would I do now? I had to find work soon or I'd lose the $500 deposit I had put down. I didn't go to bed the next morning as I usually did after night shift. I had some breakfast and started driving around, asking every place I came to if they were hiring. They say that there's one good thing about getting laid off. If you were among the *first*, you had a chance; if you were in a later bunch, all the jobs that had been available were filled with the ones who were laid off before you. (So I guess I was one of the "lucky ones!")

I can't remember how long I looked, one day or maybe three, but I did manage to find a job. It was at a sheet-metal place that made speaker systems and sound cabinets for schools. The owner's name was Johnny Walker. (And, no, he didn't make Scotch whisky!) Not only did I get a job, I got a *foreman's* job. I had to set up the machines (presses and press brakes), relegate the work to the operators (most of whom were just learning the English language), use the lathe, milling machine and do whatever welding was needed. The money was *not* good compared to Douglas Aircraft, but it was a job. I had been earning $4:40 an hour there and in this job I was getting $3:15—a little more than a dollar an hour difference; but I had to make it do.

I had been there three weeks when a young immigrant from Italy was hired. The manager came to me and asked me if I would take him under my wing and teach him some of the things I was doing. I said that I would and, even although his English was very poor, he gradually got to know what was needed in most of the jobs. Then the boss sent for me.

"Ian, I'm sorry, but I have to let you go. The young fellow you taught to do the jobs can do all right by himself now and I'm getting him for $3 an hour."

"You're letting me go for 15 cents an hour? That's only $6 a week. What are you thinking of? That young man has only a fraction of the experience that I have! You can keep your job if that's the way you think, and I'm sure you'll go out of business pretty quick with such stupid decisions." I didn't even want to finish my day's work. I told him I'd be in "tomorrow afternoon" for my wages. My new lesson was: Don't show anyone under you how to do your job; if you do, you will lose it.

So, I went looking again. The next day I got a job in Bramalea. It was in some place called "something X-Ray" and it included twelve hour shifts, constant night shift, 8 p.m. to 8 a.m. The young foreman must have been a descendant of Attila the Hun or Genghis Khan. We had been working for about two hours when I asked him what time we got our breaks.

"What do you mean 'breaks'?" he exclaimed "You get one break and that is from twenty minutes to two 'til two o'clock, and that's it."

"Just a minute, what year do you think this is? It's 1970, not 1870! You can't expect people to work for 12 hours and have only a 20-minute break."

"That's what you get. Take it or leave it."

So I took it for the time being, intending to drive down to the workplace during the day to talk to the boss.

It was approaching six a.m. and I was folding some sheet-metal on a press-break and was dead beat. That was when I decided to get a coffee from the dispensing machine and sit for a few minutes to gather together some strength. I had the coffee in my hand and my rear end hadn't even started to warm the stuff I was sitting on when the foreman flashed over to me and knocked the coffee out of my hand. (Some guys I knew back then would have punched him all over the shop and sent him to hospital, *for sure*).

"Get your f—ing ass off that lot and get f—ing working!" he shouted at the top of his lungs.

"Who the hell do you think you are talking to, you mindless idiot?" I shouted back at him, concluding, "You had better have another job lined up, Boy, because you won't have one after today if I have my way."

I immediately put all my tools away, locked my toolbox, punched the time clock, and I was on my way home.

And he *didn't* have a job later that day—I made sure of that! No one gets away with treating me like that. I can be as tenacious as a bulldog and not let go until it suits me. I was at the office in the afternoon after having a bit of sleep and related to the owner everything that had happened. I said that if that guy was being kept on, then my next step was to the Labour Board *and* the *Toronto Star* newspaper to let everyone know just how his plant operated. The boss phoned him at his home while I was with him and told him that he didn't have a job there anymore because of his handling of people. *I had fixed his wagon good!!!*

Despite this, I didn't remain there very long after that. I found a little better job and was surprised that it hadn't been taken. The rate was $3.50 an hour and it was similar work except that it was shift work, changing every week. (I found this a little better than the 12-hour shifts.) I was hoping I would be fortunate enough to remain there until I was recalled to Douglas Aircraft.

Maybe you've heard the old saying, "Dream on!" Well, I was still dreaming. The work ran out and I was laid off again.

~ * ~

What I did find out was that local employers were reluctant to hire workers from Douglas Aircraft because it was well known that they couldn't compete with the wages and conditions at Douglas. They only hired out of necessity for rush jobs, because they knew that almost without fail, the workers went back to Douglas as soon as they were recalled. This was something I had no knowledge of until I was informed of it by the foreman of the place that was just letting me go. Fortunately I was still working there when it was time to move into the new house.

~ * ~

It was only a few days before moving into our new house. I was sitting in the "wee hoose" on a Thursday afternoon when the phone rang.

"Hello, Mr. Morrans?"

"This is Ian Morrans."

"Mr. Morrans, this is Mrs. (something) from the lawyer's office to let you know that you and your wife have to come tonight at six o'clock and sign the mortgage documents."

"Oh okay, certainly, we will be there."

"Mr. Morrans, have you been made aware of the full amount required on closing."

"Well, yes. We do have the $650 balance if that's what you mean."

"Oh, I'm afraid that's nowhere near enough, Mr. Morrans. The full amount is $1,387 (plus cents)."

"No, this is wrong, surely! The realtor told us that the down payment was in two lots. We gave him the first $500 and that leaves just a balance of $650."

"Yes, that is correct, Mr. Morrans; but our fees and certain other costs have to be totaled in, along with land titles and that is the amount it comes to."

"Oh, I see." There was a sinking feeling in my stomach. The real estate agent had never mentioned that other charges had to be paid when purchasing property, nor had anyone else mentioned this to us. Of course, I was green—I didn't even know that I would have to pay the lawyer's fee! I had thought that it was included in the cost of the house, that it was part and parcel of the whole deal. I didn't even know a thing about any other costs and no one had thought of telling me fully what was involved. And there's me thinking we were doing great!

I phoned Mary at her work. "We have tae meet with the lawyer at six tonight so Ah'll meet ye down there instead of ye coming home first. Ah'll meet ye where ye come out of work."

There must have been something in my voice, for she knew that something was up. "There's something wrong, Ian; isn't there?"

"Oh, nothing much. Ah'll talk tae ye aboot it when we meet. Ye can get yersel' a cup of tea somewhere near and Ah'll meet with ye at yer office front door at around 5:45."

~ * ~

It wasn't easy telling her that we had nowhere near enough money to complete the deal. "Mary, we have to figure out exactly how much money we can scrape together by tomorrow so that I can write a cheque for this difference they want and not have it bounce. If we're not able to complete the deal, we'll lose our $500 deposit."

So there's us with pencil and paper just outside the entrance to the lawyer's office, figuring out just how much money we had. We were in danger of losing our $500 deposit and missing out on our big opportunity to better ourselves after telling everybody that we had bought a nice new house.

This was all thanks to a real estate person who didn't want to tell us of what costs were involved. I can remember leaning against some wrought iron railing to do my figuring and thinking to myself that it sure would have been nice if the initial people involved had made us wise to what was required in buying a house for the first time. I realized then that we *weren't* told by the realtor, because that would have caused us to do some quick calculations; only to find out that we couldn't make it. He knew that we were struggling to raise the balance of the down payment, so why tell us something that would lose him a sale? We just wouldn't have had the money—simple as that, so "goodbye house." I don't think he did us a favour by not telling us; it could have cost us a lot of hard-earned money which I wasn't willing to gamble with. However, I don't suppose *he* was the least bit concerned about *that*! If I hadn't been able to get a job(s), I certainly would have lost it. (To put the value into perspective, $500 was the equivalent of about four weeks' pay for me after taxes in 1970.)

(I still trusted people as I had done when we were back in Scotland; I really *should* have had more sense by then!)

We went upstairs to the lawyer's office and the papers were there ready for us to sign. All the extra costs were explained to us, down to the full tank of heating oil and taxes, etc. The young lady was very apologetic regarding our circumstances but there was nothing she could do about it. I explained that we wouldn't have enough money in the bank to cover the cheque I was writing and asked her if it could be held until the next afternoon so that I could take my pay-cheque and deposit it in the bank at noon. (Fortunately we got paid just before noon every second week and the next day was pay day.) She said that there would be no problem with that. Whew!

What we didn't tell her was that after the pay-cheque was deposited, there would be the grand sum of $5 left in our account. We had ordered carpet for the living room and runners for the stairs and hallway up to the bedrooms and the stairway down to the basement (split-level home). At the carpet outlet, I told the story, expecting them to say something about the carpet already being cut to length for us. To my relief, the owner said, "No bother, Sir. I see you are in a trade, so we'll just put it on an account for you and we'll bill you every month. How does that sound? Just sign here and you can take it with you."

Well, it sounded just great! The carpet was tied onto the top of my car and I headed for home to show Mary. "Look, Sweetheart, Ah've got the carpet for your living room."

We had never got around to applying for a credit card. It wasn't unusual for us to laugh (privately) at someone showing off a whole bunch of different credit cards strung in a plastic holder. This was completely alien to us, but we sure could have used a credit card that time.

We had got the key from the lawyer's secretary (though we didn't ever see the lawyer). The next day we took the carpet to the house and laid the main piece for the living room, so that it would be ready for the furniture to be laid on top of it.

~ * ~

I thought that our other friends would have been happy for us when we told them that we were buying a new house, but I didn't get that impression. We only saw John and Rena once after I had told them. They visited us one evening. I think the real reason was to ask me if I would put in a good word for them to the guy at the steel company, to allow them to take over the little house (that John had continually criticized since before Mary and I had moved into it.) I had to bite my lip to keep from telling him that he *didn't really want it* as it was "too small for anyone," but I didn't. I could have been catty or sarcastic, but then I figured that *they* may as well have it rather than strangers, so I told him I would most certainly see about it for him. What made his request all the more ridiculous was that they had a boy and a girl about the same ages as our two girls, and when I spoke about this he said that his son would sleep in the unfinished basement!

I arranged for them to get the house. After Mary and I moved into our new house, they moved into the little house and we never saw them again except for the house warming party they came to. It seemed to me that they were jealous of our accomplishment, and I couldn't understand it, for they had more chances than we had. And I thought they were such good friends, too.

Jack and Jean stopped visiting us also, making us wonder what it was that we had done wrong, apart from buying a house.

~ * ~

When I started this manuscript I described living in appalling poverty in Scotland. Thirty years later I found myself in a similar position because of a lying imbecile in the Government of Ontario Office. (Yes, we had a lot more "possessions" but they sure weren't paid for! We were in debt up to our eyeballs!) This book is entitled *"From Poverty to Poverty"* and I don't think that there is a more fitting description for, when coming to Canada, I certainly was right back into poverty, and worse—I had added *three* dependants!

It took us *five years* to get back to *approximately* where we were *before* we left our homeland. And those five years upward were *not* easy. I found it difficult to keep my tongue between my teeth many times in later years when someone (always a Canadian) would say to me something like, "Most likely you had to leave Scotland to get away from the poverty, the slums and the depression, to find a better life here in Canada."

You would *never* believe the number of times that a remark similar to that was said to me over the years. I'm not kidding! Sometimes it was *extremely* difficult to keep quiet, depending on how it was said, to prevent myself from telling them exactly how things were. As recently as 1998 a person from our choir was on the phone to me and said it! It took me all of my strength not to say, "Young lady, you really have *no earthly idea* what the hell you are talking about!"

Oh well, they probably wouldn't have believed me anyway, thinking that I was telling stories—making things up—for didn't everyone know that Canada was the "very best country in the world?" The States is the best place to the Americans—Germany is to the Germans, and so on. If it were otherwise, then everyone in the whole world would be living in Canada. It all depends on each individual's circumstances, doesn't it?

I was saying away back in the book that Alfred Lord Tennyson's advice had screwed up my life with the "better to have loved and lost" thing—well forget him, he turned out to be a very minor player. The guy that *really* screwed up our lives was the Government of Ontario official in Glasgow, and that's for sure. What Tennyson did was harmless; the guy in Glasgow altered our lives in such a way that it couldn't be fixed. I *do* think that we would have been inclined to settle for Australia if it hadn't been for the idea of being able to buy a new house in Canada right away.

That said, I must admit that 45 plus years after immigrating to Canada, I now consider *Canada* the best country on earth. Despite our rough start, we finally did get "established," although our financial situation was sometimes precarious. We found ourselves moving around from coast to coast and lots of places in between every two years or so. My daughters claim that they never did go to the same school two years in a row. Those years will be documented in the next volume of my autobiography entitled, "*Came To Canada, Eh?*" I guess I never did take off my "traveling shoes" for any length of time. It took a 2½ year "adventure" in Mexico in my retirement years to cure me of that and make me want to spend the rest of my life living back in Canada—but that's another leg of the story as yet unpublished.

Maybe some who read this first volume will enjoy the stories very much, maybe others won't—that's life, isn't it? I know I enjoyed writing it!

THE END

ACKNOWLEDGEMENTS

IMAGES

All maps appearing herein were drawn by the author.

All photographs were taken by the author, his acquaintances or family members or by hired professional photographers or official British military photographers as indicated.

REFERENCES

The author acknowledges the valuable assistance of *Wikipedia* in researching details on Campbeltown, Kintyre, the Suez Canal Zone history and the songs mentioned in the text.

ABOUT THE AUTHOR AND EDITOR

Ian and Gayle Moore-Morrans, pictured prior to entertaining as "Okanagan's Mr. Scotland and His Bonnie Lassie" at a Sons of Scotland Banquet, Vernon, BC, 2009.

In a unique partnership of an author and an editor who started their collaboration shortly after they were married in 2003, author Ian Moore-Morrans and his wife and editor Gayle Moore-Morrans, offer here their first non-fiction publication. They published a novel, *Beyond the Phantom Battle: Mystery at Loch Ashie*, in 2010. Ian previously had published a "how-to" e-book entitled *Metal Machining Made Easy* (under the name "Ian Morrans"), using the knowledge he had gained from years of working as a machinist.

Ian and Gayle (both widowed) met in June 2003 in Winnipeg where they started a conversation about the eclectic assortment of stories Ian had begun writing after retirement. When Ian learned that Gayle was working as Editor of *Esprit* magazine, he began to envision a future of their living and working together. They were married three months later.

After Gayle took an early retirement in July 2004, they sold their house, bought a motor home and left Winnipeg to explore retirement in Mexico. While basking in the lovely weather along Mexico's Pacific coast, Gayle started editing Ian's stories while he sat at the laptop on their RV's patio and did re-writes and touch-ups. Tiring of RV living and the hot, humid Pacific coast, they moved inland to the mountainous north shore of Lake Chapala, Mexico's largest lake. There they bought a house and became residents of the world's largest community of English-speaking ex-patriots who live in a string of small towns referred to as "Lakeside." They soon joined the Lake Chapala Society Writer's Group and met some wonderful writers from Canada, the USA, Mexico and Europe. Soon Ian's short story, "The Moonlit Meeting," was published in a local magazine, *El Ojo del Lago*.

The pair returned to Canada (but to British Columbia instead of Manitoba) in 2007, spent a year in Penticton and then moved to Vernon. They love living in the beautiful Okanagan Valley and find it perfectly suits their life-style. Despite some health challenges in the past three years, they hope to publish more of Ian's stories in the future including sequels to the novel and to this autobiography, as well as a story of revenge called "Legal Hit Man" and a number of children's stories. (Yes, it is an eclectic assortment!)

Gayle remarks, "I am also a writer but don't have the creative imagination to write fiction like Ian does. In fact, I marvel at his imagination! I also marvel at his memory in recalling all the incidents in his autobiography. The only problem was that he first wrote it using a sort-of "stream of consciousness" method. It was a challenge to sort and rearrange it, but that's what editors relish—at least this one does!"

Ian counters, "It's just as much Gayle's book as mine; in fact, I don't know what I'd do without her! I bring up the memories, analyse them and then jot them down. Then she cuts out about a third of the words, fine tunes it all (calls it "tweaking") and also acts as my agent, publicist, marketer and secretary. We're a good combination!"